Prespacetime Journal | December 2021 | Volume 12 | Issue 4

AF505120

Prespacetime Journal

Volume 12 Issue 4
December 2021

Several Physics Topics in TGD Framework

Editor:
Huping Hu, Ph.D., J.D.

Editors-at-Large:
Philip E. Gibbs, Ph.D.
Dainis Zeps, Ph.D.

<u>Advisory Board</u>

ISSN: 2153-8301   Prespacetime Journal   www.prespacetime.com
Published by QuantumDream, Inc.

(Published in Prespacetime Journal | December 2021 | Volume 12| Issue 4 | pp. 327-454)
Table of Contents

i

Table of Contents

Articles

Essay

Explorations

Article

The Role of Galois Groups in TGD Framework

Matti Pitkänen [1]

Abstract

This article was inspired by the inverse problem of Galois theory. Galois groups are realized as number theoretic symmetry groups realized physically in TGD a symmetries of space-time surfaces. Galois confinement as an analog of color confinement is proposed in TGD inspired quantum biology. Two instances of the inverse Galois problem, which are especially interesting in TGD, are following: Q1: Can a given finite group appear as Galois group over Q? The answer is not known; and Q2: Can a given finite group G appear as a Galois group over some EQ? Answer to Q2 is positive as will be found and the extensions for a given G can be explicitly constructed. The TGD based formulation based on $M^8 - H$ duality in which space-time surface in complexified M^8 are coded by polynomials with rational coefficients involves the following open question. Q: Can one allow only polynomials with coefficients in Q or should one allow also coefficients in EQs? The idea allowing to answer this question is the requirement that TGD adelic physics is able to represent all finite groups as Galois groups of Q or some EQ acting physical symmetry group. If the answer to Q1 is positive, it is enough to have polynomials with coefficients in Q. It not, then also EQs are needed as coefficient fields for polynomials to get all Galois groups. The first option would be the more elegant one. In the sequel the inverse problem is considered from the perspective of TGD. Galois groups, in particular simple Galois groups, play a fundamental role in the TGD view of cognition. The TGD based model of the genetic code involves in an essential manner the groups A_5 (icosahedron), which is the smallest simple and non-commutative group, and A_4 (tetrahedron). The identification of these groups as Galois groups leads to a more precise view about genetic code.

1   Introduction

This article was inspired by the inverse problem of Galois theory [3] (`https://cutt.ly/jmjpyDS`). Galois groups are realized as number theoretic symmetry groups realized physically in TGD [13, 14]. Galois confinement as an analog of color confinement is proposed in TGD inspired quantum biology [20, 22, 32, 19, 30, 31, 27].

Two instances of the inverse Galois problem, which are especially interesting in TGD, are following:

Q1: Can a given finite group appear as Galois group over Q? The answer is not known.

Q2: Can a given finite group G appear as a Galois group over some EQ? The answer to this question is positive as will be found and the extensions for a given G can be explicitly constructed.

The formulation adelic physics [13, 14] is based on $M^8 - H$ duality in which space-time surface in complexified M^8 are coded by polynomials with rational coefficients. Adelic physics involves the following open question.

Q: Can one allow only polynomials with coefficients in Q or should one allow also coefficients in EQs?

The idea allowing to answer this question is the requirement that TGD adelic physics is able to represent all finite groups as Galois groups of Q or some EQ acting physical symmetry group.

If the answer to **Q1** is positive, it is enough to have polynomials with coefficients in Q. It not, then also EQs are needed as coefficient fields for polynomials to get all Galois groups. Needless to say, the first option would be the more elegant one.

In the sequel the inverse problem is considered from the perspective of TGD.

---

[1] Correspondence: Matti Pitkänen `http://tgdtheory.fi/`. Address: Rinnekatu 2-4 8A, 03620, Karkkila, Finland. Email: matpitka6@gmail.com.

1. $M^8 - H$-duality, $H = M^4 \times CP_2$ adelic physics [13, 14] based on the identification of space-time surfaces X^4 in the complexified M^8 identifiable as complexified octonions. Normal space of X^4 is required to be associative/quaternionic and to contain an integrable distribution of commutative 2-D sub-spaces. At the level of H, twistor lift of TGD implies partial differential equations defined by a variational principle based on action, which is sum of volume term and Kähler action.

   The spacetime surfaces are preferred extremals [25, 28] identifiable as minimal surfaces analogous to soap films except at the dynamically generated analogs of frames at which the minimal surface property fails and field equations hold true only for the full action. At the frame, the divergences of isometry currents for the volume term and Kähler action have delta function singularities which cancel each other.

   In M^8, space-time surfaces are determined as 4-D "roots" of polynomials of complex variable over rationals continued to an octonionic polynomial [16, 17, 23, 28]. The Galois group over Q acts as a physical symmetry group permuting the sheets of the X^4 defined by the 4-D roots. This action induces symmetry also at the H-side of the duality and the Galois group defines a new kind of symmetry, which distinguishes between TGD and competing theories.

2. The notion of infinite primes is inspired by TGD [7] and has a physical interpretation as a repeated second quantization of an arithmetic super-symmetric quantum field theory (QFT). This hierarchy corresponds to a hierarchy of multi-variable polynomials obtained by taking a polynomial of t_n and replacing its rational coefficients with rational functions of $t_1, ..t_{n-1}$ with rational coefficients. One can assign a Galois group to these polynomials and therefore also to infinite primes and integers.

3. The fraction of 2-groups with order not larger than n approaches unity at the limit $n \to \infty$. All 2-groups act as Galois groups of space-time surfaces. p-Adic length scale hypothesis states that primes near power of 2, define physically preferred p-adic length scales. The special role of 2-groups might explain why the p-adic length scale hypothesis [5] is true.

4. Galois groups, in particular simple Galois groups acting on cognitive representations consisting of points, whose coordinates in a number theoretically preferred coordinate system of octonions belong to EQ, play a fundamental role in the TGD view of cognition [21]. The TGD based model of genetic code [8, 20] involves in an essential manner the groups A_5 (icosahedron (I)), which is the smallest simple and non-commutative group, and A_4 (tetrahedron (T)). Genetic code has as building bricks Hamiltonian cycles of I and T. Genetic code relates to information and therefore to cognition so that the interpretation of these symmetry groups as Galois groups is suggestive.

   The most recent step of progress was the realization that genetic code can be represented in terms of icosa-tetrahedral tesselation of a hyperbolic 3-space H^3 [22] and that the notion of genetic code generalizes dramatically. Also octahedron (O) is involved with the tesselation but plays a completely passive role. The question why the genetic code is a fusion of 3 icosahedral codes and of only a single tetrahedral code remained however poorly understood.

   The identification of the symmetry groups of the I, O, and T as Galois groups makes it possible to answer this question. Icosa-tetrahedral tesselation can be replaced with its 3-fold covering replacing $I/O/T$ with the corresponding symmetry group acting as a Galois group. T has only only a single Hamiltonian cycle and its 3-fold covering behaves as a single cycle.

2   Some background about Galois groups

2.1   Basic definitions

Galois extensions are by definition represented by the roots of polynomials with coefficients in K. By definition Galois group Gal leaves K invariant and permutes roots. For instance, complex conjugate roots are permuted.

ISSN: 2153-8301   Prespacetime Journal   www.prespacetime.com
Published by QuantumDream, Inc.

There are two basic ways to construct Galois extensions L/K of a number field K.

1. The roots of irreducible polynomials over K (no rational roots in K) define a Galois extension. The order $ord(Gal)$ of Gal is equal to the dimension n of extension L: $ord(Gal) = n$.

2. If a number field L and its automorphism group Aut is known then any subgroup G of Aut defines a sub-field K^G invariant under G and L/K^G an extension having G as Galois group.

The functional composition $P_1 \cdot P_2$ creates an extensions for which the Galois group of P_2 is a normal subgroup if one has $P_i(0) = 0$ and $P_1 \circ P_2$ has also the roots of P_2.
Polynomial rings $K(t_1, ..., t_n)$ of several variables give rise to extensions via the roots $P(t) = 0$.

1. The permutation group S_n acts as automorphisms of $K(t_1, ...t_n)$. Especially interesting sub-field is the invariant field of S_n generated by polynomials generated by symmetric functions. Any finite group G is sub-group of some S_n and defines a G-invariant field of $K(t_1, ...t_n)$ as G-invariant rational functions containing the field generated by symmetric functions.

2. The completion of a number field is algebraically complete. For instance, for Q this completion $\overline{Q}$ consists of algebraic numbers. The Galois group of $\overline{Q}/Q$ is profinite, which means that it is infinite but effectively finite, and can be constructed by inverse limit construction for a sequence of extensions leading to algebraic numbers.

   The extensions of completions are necessarily transcendental. In the case of $\overline{Q}$ they involve addition of transcendentals the extensions.

2.2   Some results about Galois groups over rationals

It is good to start by listing some basic results related to the Galois groups [3].

1. Galois theorem states that polynomials are solvable for degree $d \leq 5$. In these cases the Galois group is solvable meaning EQ is extensions of extension .... of rationals and that Galois groups for EQ has a descending decomposition by normal groups H_i which are commutator groups of the normal group H_{i+1} at the previous level. Equivalently, the Galois group for an extension of an extension at level i is Abelian. For $d < 5$ Galois group is $A_5 = S_5/Z_2$ with 60 elements. This is the smallest non-Abelian simple group. By its definition, a simple group does not have a decomposition to normal groups.

2. Kronecker-Webber theorem states that any abelian group appears as a Galois group for Q. This was found by studying cyclic extensions.

3. Shafarevich proved that every solvable group appears as a Galois group for an EQ.

4. Scholz and Reichardt proved that for an odd prime p, every finite p-group occurs as a Galois group over Q. The order of each element of a simple group is a power of p. 2-groups which also appear as a Galois group over Q are of special interest since for given n, most groups with order smaller than n are 2-groups. This result is of special interest from the point of view of p-adic length scale hypothesis.

5. It has been conjectured that almost all finite groups can act as a Galois group over Q. (`https://cutt.ly/imjwDKC`).

6. Simple groups are primes for finite groups. Simple groups appear in the decomposition of EQ to a sequence of extensions with a simple Galois group represented by a hierarchy of polynomials. In TGD inspired theory of cognition simple groups are analogs of elementary particles [21, 26, 33, 23] so that they are of special interest.

Any finite group, in particular, any simple group, appears as a Galois group over Q. The open question is whether a given simple group can appear as a Galois group over Q.

Many simple groups appear as Galois groups over Q. The theorem of Malle and Matzat states that if p is an odd prime such that either 2, 3 or 7 is a quadratic non-residue modulo p (q is quadratic residue if $x^2 = q \ mod \ p$ has a solution) then $PSL(2,p)$ occurs as a Galois group of EQ.

Four of the Mathieu groups, namely M_{11}, M_{12}, M_{22} and M_{24}, occur as Galois groups of EQ. For M_{23} the situation is open. The theorem of Thompson states that the Monster group, the largest sporadic simple group, appears as a Galois group over Q.

2.3 Various problems related to inverse Galois problem

In [3] various problems related to inverse Galois problem are listed.

1. The general existence problem can be formulated as the following question. Given number field K and finite group G, can G act as a Galois group for some extension of K?

2. If the answer to the general existence problem is affirmative for given K and G, the explicit construction polynomials $P(t)$ is the next challenge.

   One can also consider polynomials $P(t_1, ..., t_n)$ of several arguments which could be regarded as parametric representation for a large number of polynomials. $P(t_1, ..., t_n)$ must be irreducible. If the restriction of arguments appearing as parameters to a specific value produces irreducible polynomial, one can hope that the Galois group over Q is same as that of the polynomial which is permutation group S_n of the arguments.

   This process is called specification, and Hilbert's irreducibility theorem states the conditions for when the irreducibility is preserved in the process so that Galois group is inherited. The conditions mean that the Galois group is almost independent on the parameter values and the loci were irreducibility fails are the places where this happens. Obviously they correspond to the occurrence of multiple roots. Hilbert proved that for S_n the conditions are satisfied so that they appear as Galois groups of some EQ.

   Also the subgroups $G \subset S_n$ act in $Q(t_1, ...t_n)$ and one can ask under what conditions one can find EQ for which G acts as a Galois group. It turns out that one can construct explicty an EQ, whose extension allows G as a Galois group and specify explicitly the conditions under which this EQ reduces to Q.

3. What is the smallest number of parameters for a generic polynomial $P(t_1, ...t_n)$?

Two special results are mentioned in [3]. G can be any finite group in the following cases involving only one parameter.

1. For $K = C(t)$ any finite group G appears as Galois group of some Galois extension (defined by polynomial) of $K = C(t)$. This is true also for Galois extensions of generalizes to $K = R(t)$ and $K = \overline{Q}(t)$.

   The result for $K = C(t)$ follows from Riemann existence theorem `https://cutt.ly/FmjsnPA`, which in its original form states that the space of functions on Riemann sphere having singularities at punctures can be regarded as space of analytic functions at Riemann surface obtained as a finite branched covering of S^2 with branchings at punctures.

   The absolute Galois group over $C(t)$ corresponding to an infinite covering of the puncture sphere is identifiable as homotopy group of infinite covering and is free profinite (infinite but effectively finite) group with infinite number of generators. This true when K is closed and therefore holds true for $\overline{Q}$. Extension of $K(t)$ is obtained by adding a parameter and finding the roots.

ISSN: 2153-8301

www.prespacetime.com

The homotopy group of a given finite covering corresponds to a braid group as a finite covering of S_n. This raises the idea that a given finite group having always a representation as subgroup of S_n could allow a construction giving G as Galois group of over Q.

2. If K is p-adic field Q_p, any finite group can act as a Galois group over $K = Q_p(t)$.

3   Methods

In this section various methods to answer to the question whether a given finite group can act as a Galois group for given number field K are briefly summarize. The discussion follows the discussion in [3].

3.1   Regular Inverse Galois Problem

Regular inverse Galois problem starts from an extension L/K, which is regular.

Regularity requires that K is algebraically closed in L - or L is purely transcendental extension of K. This means that the elements of K cannot be expressed as solutions of algebraic extensions in L. One example of a purely transcendental extension is extension of rationals by adding some transcendental numbers. If K is algebraically closed - this is the case for C and algebraic numbers - the condition is satisfied.

A further condition is that L is separable over K. For a physicist, this rather technical looking condition states that the number field $L \otimes_K \overline{K}$ is an integral domain meaning that it has no divisors of zero.

Transcendental extensions are regular. The so-called transcendence basis S consists of elements of L, which do not satisfy any algebraic equation in L. One can construct the field $K(S)$ by forming the product of basis for K and S and $L/K(S))$ is algebraic extension of $K(S)$.

Extensions of algebraic completions are regular/transcendental. The field defined by rational functions formed from S_n invariant polynomials in $K(t_1, ..., t_n)$ define field K^G of symmetric rational functions which cannot be regarded as an algebraic extension of K. By Noether's theorem stating that K^G is isomorphic with $K(t_1, ..., t_n)$, $K(t_1, ..., t_n)/K^G$ defines an extension of K with the same Galois group.

Whenever one has a Galois extension $M/Q(t)$ (regular or not), it is an easy consequence of the Hilbert Irreducibility Theorem that there is a specialisation M/Q with the same Galois group. If $M/Q(t)$ is regular, one obtains such specialized extensions M/K over any Hilbertian field in $char = 0$, in particular over all algebraic number fields. Char denotes the integer n for which $nx = 0$ is true for all elements of the field. Finite fields F_p have $char = p$.

$C(t)$ and $\overline{Q}(t)$ are algebraically closed and transcendental and any finite group defines a Galois group for the extensions of these fields are obtained from a polynomial of n variables by specification.

The problem is how to get down to $Q(t)$ from $\overline{Q}$. One must restrict the coefficients of polynomials to a sub-field and it is not clear what happens to the Galois group in this process. This is one case of specialization: one starts from a parametrized set of polynomials $C(t_1, ...t_n)$ or $\overline{Q}(t_1, ...t_n)$ and restricts the parameters $t_1, ..., t_{n-1}$ to say Q.

3.2   Hilbert's irreducibility theorem

Consider polynomials $f(t, x)$ with parameters $t = (t_1, ...t_r)$ and indeterminates $x = (x_1, ...x_n)$. Assume that f irreducible polynomial. Define Hilbert f-set H_f/K as the set of parameter values $t \in K^r$ for which the restriction is well-defined and irreducible. Define Hilbert g-set as a subset of the set of the parameter values $t \in K^r$ for which $g(t)$ is non-vanishing so that g has no zeros in the set defined by the points $(x_1, ...x_n)$: this is possible since K^r is a subset of all parameter values t. Define Hilbert set as an intersection of finitely many f-sets and finitely many g-sets.

The field K is said to be Hilbertian if Hilbert sets are nonempty for all r.

The above condition is rather abstract but the following characterization of Hilberianity is more concrete. For a field K with $char = 0$, K is Hilbertian if and only if the following condition holds true. If $f(t, X)$ has no roots in $K(t)$ then $f(a, X)$ has no roots in K.

For $K = Q$, the first condition means that there are no roots which are rational functions and the second condition means that $f(a, X)$ has no rational roots. Rational roots emerge when two or more algebraic roots coincide. In this situation, the irreducibility is preserved in the specification and the Galois group is inherited.

Hilbert also proved that for S_n acting as a Galois group for $Q(t_1, ..., t_n)$, it is possible to find an extension of rationals with the same Galois group by specification. The polynomials in question are invariant under S_n and generated by symmetric functions. If the specification has a root, the action of S_n to a root corresponds to the action of a permutation in the parameter space and creates a new root so that the Galois group is S_n.

Under the conditions stated by Noether, this generalizes to subgroups $G \subset S_n$.

3.3 Noether's problem

Algebraic numbers are algebraically complete and can have only transcendental extensions, say by addition of transcendental numbers.

On the case of polynomial algebra $Q(t_1, ..., t_n)$ the field of invariants Q^G, $G = S_n$ is generated by polynomials symmetric under permutations of n arguments acting as Galois group in the polynomial algebra $C(t_1, ...t_n) \equiv C(t)$. Q^{S_n} is transcendental in the sense that the generators do not satisfy polynomial conditions with coefficients in $C(t)$. This algebra has $C(t)$ as extension with Galois group S_n, which obvious commutes with the field operations.

One can consider also sub-groups of $G \subset S_n$ and analogous extensions. In this case it is not obvious that the algebra Q^G is rational which meas that $C(t)$ is purely transcendental extension of Q^G.

The theorem by Emmy Noether states the following: If G is finite and $Q(X)^G/Q$ is rational (purely transcendental), then there is a Galois field extension K/Q with group G.

The proof of the theorem involves Hilbert's irreducibility theorem, rationality property implying that Q^G is isomorphic with $Q(t_1, ...t_n)$, and primitive element theorem stating that the extensions of Q are generated by powers of primitive element. How G becomes the Galois group for Q has been already explained.

3.4 Rigidity method

Riemann existence theorem is an essential part of the rigidity method. One considers compactified plane with punctures allowing interpretation as a punctured sphere with origin as a marked point. Rieman proved that meromorphic functions singular at punctures functions can be regarded as regular functions in cover of S^2 branched at the punctures defining a Riemann surface.

The homotopy group of the sphere with n punctures has n generators $g_1, ..., g_n$ satisfying the relation $g_1...g_n = 1$, since the complement of the regions containing punctures contains no punctures.

There is an infinite number of coverings characterized by the number n of sheets. Intuitively they are analogous to functions $z^{1/n}$. The homotopy group gives rise to the homotopy group of n-fold covering acting also as a Galois group for extension of meromorphic functions induced by the cover. The Galois group serves also as a braid group defining n-fold covering group for the permutation group of the punctures.

Absolute Galois group is associated with the covering with $n = \infty$ and is pro-finite group (infinite but effectively finite). Also this group satisfies the analog of the relation $g_1...g_n = 1$. The absolute Galois group is obtained as an inverse limit of the groups associated with a sequence of extensions (`https://en.wikipedia.org/wiki/Inverse_limit`). The map h_{jk} from k:th level to $j \leq k$:th level is homomorphism but not isomorphism for $k > j$ because it is many-to-one. h_{ii} is identity homomorphism. Compatibility condition $h_{ik} = hijhjk$ is satisfied.

Since the braid group B_n is a covering of S_n, any finite group is sub-group of some B_n. Therefore subgroups of B_n could act as Galois group for Q under suitable conditions. Note that braid groups as coverings of S_n also relate to quantum groups and are therefore physically highly interesting.

According to [3], a considerable progress has been made in the realization of simple groups as Galois groups of regular extensions over $C(t)$ and, and by Hilbert's irreducibility theorem, over every number field.

The basic idea of the rigidity method is that every finite group is a Galois group of some covering of a polynomial field with coefficient in $C(t)$ and $\overline{Q}(t)$. What covering means that one has effectively many-valued polynomials (recall the analog with function $z^{1/n}$) and Galois group permutes the values at a given point. One must only identify the conditions, which ensure that the polynomial can be defined over $Q(t)$.

Rigidity method helps to get down to Q. It is shown that there exists an extension with a given finite Galois group G over some EQ. EQ is generated by the values of G characters for $r \geq 3$ classes of G. Already this is an interesting result. However, if the characters are rational valued, EQ reduces to that for Q and has G as Galois group permuting the copies of the many-valued function.

4   Connections with TGD

How Galois groups emerge in TGD framework, was discussed in the introduction. In this section the connections of the inverse Galois problem with TGD are discussed.

4.1   Why the inverse Galois problem is so relevant for TGD?

The formulation of TGD relies on $M^8 - H$ duality in which space-time surface in complexified M^8 with octonionic interpretation are coded by polynomials with rational coefficients involves the following open question.

Q: Can one allow only polynomials with coefficients in Q or should one allow also coefficients in EQs?

The condition allowing to answer this question is that adelic physics [13, 14] must be able to represent all finite groups as Galois groups over Q or some EQ as physical symmetry group. More generally, TGD Universe is able to physically represent all internally consistent mathematics.

1. If any finite groups can serve as a Galois group over Q, it is enough to have polynomials with coefficients in Q.

2. It this is not possible, then also EQs are needed as coefficient fields for polynomials to have all possible finite groups as Galois groups. The answer to this question is positive. One studies rational functions of a complex variable in S^2 having singularities at n punctures. The Galois group is identified as a braid group for n braids identifiable as a covering group of S_n. G is identified as a subgroup of the braid group.

   One constructs first an extension of certain EQ with Galois group G. As already explained, EQ can be explicitly constructed in terms of the characters of G assignable to $r \geq 3$ conjugacy classes of G and defined as traces of the matrices representing the group element. G acts as a conjugation. This extension of EQ has G as a Galois group. If the characters are rational, the extension is trivial and G acts as a Galois group over Q.

Needless to say, the first option would be the more elegant one.

The folklore is that the inverse Galois hypothesis is true for very many simple groups (this is true at least in the sense that almost all simple groups are 2-groups). Simple groups do not have a non-trivial normal subgroup decomposition so that the polynomial defining the extensions is not representable as a functional composite of polynomials. If all simple Galois groups can appear as Galois groups over rationals then extensions with non-simple Galois group could correspond to composite polynomials.

Note that the functional composition of polynomials yields a fractal structure at space-time level. The polynomial P_{n_1} resp. P_{n_2} corresponds to n_1- resp. n_2-sheeted and $P_{n_1} \cdot P_{n_2}$ corresponds to n_1-sheeted structure with each sheet consisting of n_2 sheets. The question whether functional iteration of a polynomial P_n could define an analog for the approach to chaos at space-time level in the sense of Mandelbrot and Julia fractals is discussed in [18]. Also the functional composites involving different polynomials P_{n_i} should lead to fractal-like structures at the space-time level.

For $P(0) = 0$, the roots of polynomial P are possessed also by its iterate and one could glue regions defined by m:th iterate and time reversal of n:th iterate at values $t = t_n$ corresponding to the roots of P to get a a sequence of iterates with various values of n [28]. In this case the roots are conserved and this brings in mind the notion of conserved genes. It is difficult to avoid the idea that genes could at the level of the magnetic body of the gene actually correspond to functional composites of polynomials P_i satisfying $P_i(0) = 0$.

4.2   Galois invariance as a physical symmetry in TGD

Adelic physics [12, 13] is a proposal for the physics of both sensory experience having real physics as correlate and cognition having various p-adic physics as correlates. Adele is a book-like structure formed by real numbers and the extensions of p-adic number fields induced by a given extension of rationals with the pages of the book glued together along its back consisting of numbers belonging to the extension of rationals. This picture generalizes to space-time level. Adelic physics relies on the notion of cognitive representation as a unique number theoretic discretization of the space-time surface. This discretization has also fermionic analog in terms of spinor structure associated with the group algebra of the Galois group over Q.

4.2.1   Adelic physics very briefly

Number theoretic vision leading to adelic physics [14] provides a general formulation of TGD complementary to the vision [6] (`http://tinyurl.com/sh42dc2`) about physics as geometry of world of classical words (WCW).

1. p-Adic number fields and p-adic space-time sheets serve as correlates of cognition. Adele is a Cartesian product of reals and extensions of all p-adic number fields induced by given extension of rationals. Adeles are thus labelled by extensions of rationals, and one has an evolutionary hierarchy labelled by these extensions. The large the extension, the more complex the extension which can be regarded as $n - D$ space in K sense, that is with K-valued coordinates.

2. Evolution is assigned with the increase of algebraic complexity occurring in statistical sense in BSFRs, and possibly also during the time evolution by unitary evolutions and SSFRs following them. Indeed, in [18] (`http://tinyurl.com/quofttl`) I considered the possibility that the time evolution of self in this manner could be induced by an iteration of polynomials - at least in approximate sense. Iteration is a universal manner to produce fractals as Julia sets and this would lead to the emergence of Mandelbrot and Julia fractals and their 4-D generalizations. In the sequel will represent and argument that the evolution as iterations could hold true in exact sense.

   Cognitive representations are identified as intersection of reality and various p-adicities (cognition). At space-time level they consist of points of imbedding space $H = M^4 \times CP_2$ or M^8 ($M^8 - H$ duality [9, 10, 11] allows to consider both as imbedding space) having preferred coordinates - M^8 indeed has almost unique linear M^8 coordinates for a given octonion structure.

3. Given extension of given number field K (rationals or extension of rationals) is characterized by its Galois group leaving K - say rationals - invariant and mapping products to products and sums to sums. Given extension E of rationals decomposes to extension E_N of extension E_{N-1} of ... of extension E_1 - denote it by $E \equiv H_N = E_N \circ E_{N-1}... \circ E_1$. It is represented at the level of classical

space-time dynamics in M^8 (http://tinyurl.com/quofttl) by a polynomial P which is functional composite $P = P_N \circ P_{N-1} \circ ... \circ P_1$. with $P_i(0) = 0$. The Galois group of $G(E)$ has the Galois group $H_{N-1} = G(E_{N-1} \circ ... \circ E_1)$ as a normal subgroup so that $G(E)/H_{N-1}$ is group.

The elements of $G(E)$ allow a decomposition to a product $g = h_{N-1} \times h_{N-1} \times ...$ and the order of $G(E)$ is given as the product of orders of H_k: $n = n_0 \times .. \times n_{N-1}$. This factorization of prime importance also from quantum point of view. Galois groups with prime order do not allow this decomposition and the maximal decomposition and are actually cyclic groups Z_p of prime order so that primes appear also in this manner.

Second manner for primes to appear is as ramified primes p_{ram} of extension for which the p-adic dynamics is critical in a well-defined sense since the irreducible polynomial with rational coefficients defining the extension becomes reducible (decomposes into a product) in order $O(p) = 0$. The p-adic primes assigned to elementary particles in p-adic calculation have been identified as ramified primes but also the primes labelling prime extensions possess properties making them candidates for p-adic primes.

Iterations correspond to the sequence $H_k = G_0^{\circ k}$ of powers of generating Galois groups for the extension of K serving as a starting point. The order of H_k is the power n_0^k of integer $n_0 = \prod p_{0i}^{k_i}$. Now new primes emerges in the decomposition of n_0. Evolution by iteration is analogous to a unitary evolution as ex^{iHt} power of Hamiltonian, where t parameter takes the role of k .

4. The complexity of extension is characterized by the orders n and the orders n_k as also the number N of the factors. In the case of iterations of extension the limit of large N gives fractal.

5. At space-time level, Galois group acts in the space of cognitive representations and for Galois extensions for which Galois group has same order as extensions, it is natural do consider quantum states as wave functions in $G(E)$ forming n-D group algebra. Therefore Galois groups becomes physical symmetry groups.

   One can assign to the group algebra also spinor structure giving rise to $D = 2^{M/2}$ fermionic states where one has $N = 2M$ or $N = 2M + 1$). One can also consider chirality constraints reducing D by a power of 2. An attractive idea is that this spinor structure represents many-fermion states consisting of $M/2$ fermion modes and providing representation of the fermionic Fock space in finite measurement resolution.

Adelic physics [14], $M^8 - H$ duality [16, 17, 23, 28], and zero energy ontology lead (ZEO) [15, 25, 24] to a proposal [21] that the dynamics involved with small state function reductions (SSFRs) as counterparts of weak measurements could be basically number theoretical dynamics with SSFRs identified as reduction cascades leading to completely un-entangled state in the space of wave functions in Galois group of EQ identifiable as wave functions in the space of cognitive representations. As a side product a prime factorization of the Galois group to simple factors as normal subgroups is obtained.

The result looks even more fascinating if the cognitive dynamics is a representation for the dynamics in real degrees of freedom in finite resolution characterized by the extension of rationals. If cognitive representations represent reality approximately, this indeed looks very natural and would provide an analog for adele formula expressing the norm of a rational as the inverse of the product of is p-adic norms. The results can be appplied to the TGD inspired model of genetic code.

4.2.2 The notions of invariant field and Galois confinement

The physical meaning of the invariant field is interesting from the TGD point of view. I have proposed the notion of Galois confinement as a generalization of color confinement stating that physical states are invariant under Galois group. Color confinement would force hadrons to behave like single unit so that one cannot observe free quarks and the same would happen now.

ISSN: 2153-8301   Prespacetime Journal   www.prespacetime.com
                            Published by QuantumDream, Inc.

For instance, in living systems the states of magnetic bodies could be Galois singlets with respect to appropriate Galois group. This would guarantee their stability. For istance, units consisting of 3 dark protons ($h_{eff} = nh_0 > h$) and of 3 dark phonons would represent genetic codons. Galois confinement would force them to behave like single unit. In double DNA strand the codon and is conjugate would form this kind of pair. Dark $3N$-protons and $3N$-photons would in turn represent genes [32, 30, 31, 20, 22].

Galois invariance means invariance under permutations of space-time sheets by the action of Gal and also the invariance of many-fermion states proposed to correspond to the $2^{[n/2]}$-D spinor space for the spinors assignable to the $n - D$ extension having interpretation as fermionic Fock states.

Galois confinement cannot be permanent. In transitions changing the value of h_{eff} it could be lost. For instance, gene can decay to codons and DNA strand could split during replications and transcription.

4.3   The physical interpretation of multi-variable polynomial rings in TGD

Consider the polynomial ring $Q(t_1, .., t_n)$ over Q defining a field of rational functions. This set could define a parametrized set of space-time surfaces. By solving the roots with respect to t_n by keeping $t_1, ...t_{n-1}$ as parameters, one obtains some number of roots. This gives rise to an extension of $Q(t_1, ..., t_{n-1})$ involving algebraic functions of $t_1, ..., t_{n-1}$. One can actually solve the roots of the polynomial equation with respect to any variable t_k. The degrees of the polynomial with respect to t_i are in general different and this would mean that the orders of the Galois group are different.

Here the Noether's theorem comes to the rescue. Instead of polynomials in $Q(t_1, ...t_n)$, one can consider symmetrized polynomials invariant under S_n for which the degree is same for all variables t_i and Galois groups have the same order. Therefore the specification with respect to any t_i can give rise to an irreducible polynomial with the same Galois group. In this case, the extensions defined by the roots of polynomials give also an extension of Q with the same Galois group except at points, where the restriction fails to be irreducible.

Noether's theorem considers a situation for a subgroup of $G \subset S_n$. If the G-invariant field Q^G is purely transcendental and therefore isometric with K, the Galois group of extension of Q^G defines a Galois group over Q. Hilbert's theorem states that this is the case for $G = S_n$.

The specification obtained by fixing the values of $t_1, ..., t_{n-1}$ must be irreducible. The Galois group for the restriction of the polynomials in Q^G is the same for all parameter values at which irreducibility is true.

4.3.1   Could multi-variable polynomials define sub-spaces of WCW with a given Galois group

Space-time surfaces are determined by polynomials of a complex variable with rational coefficients by algebraically continuing them to polynomials of a complexified octonion.

Polynomials with several variables play a central role in the theory of Galois groups. In the TGD framework the parameter type variables would give a parameterized set of space-time surfaces with the same Galois group except at the points at which the irreducibility fails. The order of parameters matters unless one considers only polynomials invariant under S_n or in some cases its sub-group G and inverse Galois theorem does not hold true as is clear from the fact that the dimension of the local Galois group depends on what variable one regards as the variable which is solved.

This kind of parameter sets would naturally define sub-WCW) (WCW is shorthand for "the world of classical worlds") and allow to define WCW spinor fields defining quantum superpositions of space-time surfaces with the same Galois group except at a the points at which the irreducibility of the restriction to a polynomial of a single variable fails.

At the level of sub-WCW, Galois invariance would mean a restriction to the S_n invariant field of a polynomial ring defined by symmetrized multi-variable polynomials gives a parametrized set of extensions of rationals for the "behavior" variable as a complex coordinate continued to complexified octonionic coordinate of M^8. One can also consider the restricted symmetry defined by $G \subset S_n$ encountered in Noether's theorem.

1. If the Galois group is S_n, a possible physical interpretation of Galois confinement would be as a realization of Bose-Einstein statistics in the bosonic degrees of freedom of WCW. Also fermionic statistics could allow a similar interpretation.

2. Could the Galois confinement with respect to a subgroup $G \subset S_n$ have an interpretation in terms of anyonic statistics and charge fractionalization? Could the condition that the invariant field defines a transcendental extension and hence G acts as a Galois group over Q serve as a physical constraint.

3. On the other hand, the fact that anyonic statistics is essentially a 2-D phenomenon associated with the braid group suggests that it could be assigned to the function field $Q(z)$ at a partonic 2-surface containing fermions as punctures. In $M^8 - H$ duality, the positions of fermions would have interpretation as singularities and their position would represent WCW coordinates of the space-time surface. If the strongest form of holography holds true, the position of these punctures could code for the space-time surface (real polynomials are determined by their values at a finite number of points).

4.3.2 The failure of specification, catastrophe theory, and quantum criticality

In Hilbert's irreducibility theorem the notion of specification is essential. Specification fails when it produces as a restriction a reducible polynomial decomposing into a product of polynomials.

1. From the factorization in terms of roots it follows that this occurs when 2 or more roots coincide. For instance, if the roots correspond to a conjugate pair of real or complex roots, they become degenerate and rational. In this case the order of the Galois group decreases.

2. The polynomial can decompose to a more general product meaning a decomposition of the Galois group to a product. One can imagine that there is a small term added to a product of polynomials which vanishes at the criticality. The corresponding space-time region decomposes to distinct regions which can intersect at discrete points. Note for rational polynomials the critical situation is not achieved by a smooth change of the parameters.

   Geometrically criticality means that the space-time surface decomposes to disjoint surfaces corresponding to the roots of the factors which define lower-D extensions of rationals with smaller Galois groups. The decay of the space-time surface occurs. Particle reactions in the geometric sense could correspond to this kind of critical situation.

For the first alternative, catastrophe theoretic analogy [2] (`https://cutt.ly/9mEG8gn`) helps to gain some physical intuition. In the simplest situations such as cusp catastrophe, one has one behavior variable x and some number of control parameters t_i. The roots are those of the gradient of the dV/dx. The equation $dV/dx = 0$ for equilibrium states gives rise to a catastrophe graph in the space defined by x and control parameters. One restricts to real roots so that the map decomposes to regions characterized by different numbers of real roots.

The cusp catastrophe is a simple example. In this case dV/dx has degree $d = 3$ allowing 3 real roots or 1 real root and complex conjugate pair or roots. By restricting x to be real, these correspond to regions which are 3 sheeted and 1-sheeted covers of the 2-D parameter space. At the different sides of the boundary two real *resp.* complex roots become degenerate and irreducibility fails.

This situation corresponds to a criticality at which sudden catastrophic changes can occur. Therefore also the lower-dimensional regions where irreducibility fails are physically highly interesting.

Self-organized criticality (SOC) (`https://cutt.ly/xmEHgcN`) is a real phenomenon but very difficult to understand in thermodynamics with a single arrow of time and should also have a number theoretical interpretation. Zero energy ontology (ZEO) [15, 25] is crucial for the formulation of quantum measurement theory in the TGD framework. This theory extends to a theory of consciousness and leads to a model of self-organized quantum criticality (SOQC) [32, 29, 24] .

One of the key predictions is that the arrow of time changes in the TGD counterparts of ordinary state function reductions (SFRs) - "big" SFRs (BSFRs). For the non-standard arrow of time, dissipative processes look like self-organization processes. This leads to an understanding of (SQOC). The state of the critical sub-system S_1 is unstable but in a time direction opposite to the arrow of time for the system (S). Hence the S_1 tends to criticality when viewed by S. The critical surfaces of the parameter space correspond to the analogs of catastrophes as a failure of reducibility: life seems to love catastrophes!

4.3.3   Hierarchies of parametrized polynomials and of infinite primes

The discovery of the hierarchy of infinite primes and their correspondence with a hierarchy obtained by a repeated second quantization of arithmetic quantum field theory gave a strong boost for the speculations about TGD as a generalized number theory [7].

1. At the lowest level, ordinary primes p label bosonic and fermionic states of an arithmetic super-symmetric quantum field theory (QFT). The product $X_1 = \prod_p p$ of all finite primes is infinite as a real number but finite as a p-adic number for all primes p (the norm is $1/p$) and can be regarded as an analog of the Dirac sea.

   Infinite primes (having unit p-adic norm for every p) are created by kicking from the Dirac sea a set of negative energy fermions represented by the product $n = \prod_{k \in U} p_k$ primes p_k. This gives rise to an object $P = X_1/n + n$. It is easy to check that P is prime. One can also add bosons by the replacement $P \to kX_1/n + ln$ such that k does not divide n and l has a decomposition to primes appearing as prime factors of n. Altogher $m+l$ bosons have been added.

   The simplest infinite primes are linear in the formal variable X_1 as analogs of roots of monomials with rational coefficients and analogous to Fock states of free bosons and fermions. Besides analogs of Fock states, also analogs of bound states are obtained as infinite primes. They correspond to irreducible polynomials $P(X_1)$ of a single variable obtained. One can decompose P just like an ordinary polynomial to factors corresponding to the roots of $P(X_1) = 0$.

   Infinite primes have therefore an interpretation as many-particle states of a supersymmetric QFT with bound states included and represented in terms of extensions of rationals. There is no need to emphasize that bound states represent a basic problem of QFTs.

2. At the next level of the hierarchy infinite primes X_1 is replaced with X_2 as a product of infinite primes at the first level of the hierarchy and the construction can be continued. The formal variables X_i characterizing various levels of the hierarchy correspond to the infinite (in real sense) numbers defined by the products of all primes at the previous level.

   It is possible to decompose the polynomials at level n to products of monomials defined by the roots of the polynomial equation with X_n as an independent variable to be solved. The infinite primes at the first level become single particle states and second quantization is repeated. At n:th level, one can also construct irreducible polynomials of $X_1, X_2, ..X_n$ and obtains analogs of bound states. One can decompose these polynomials just like one decomposes polynomials of ordinary variables x_i. This gives rise to algebraic extension associated with the rational field defined by polynomials of $X_1, ..., X_n$.

3. The polynomials associated with infinite primes are ordered. The polynomial at n_{th} level has polynomials at $n-1$:th level as coefficients. The nature of the construction as a hierarchy of second quantizations gives rise to states formed from states formed from ...

   This suggests that the symmetrization with respect to variables X_i leading to S_n invariant field realized in terms of symmetric functions does not make sense physically. Notice that the polynomial obtained by the symmetrization does not represent an infinite prime. Physically the different levels in the quantization could correspond to space-time surfaces, whose size scales increase with n. Space-time surfaces at level $n-1$ would be glued by wormhole contacts to the space-time surfaces at the level n.

4. In M^8 picture, infinite primes mapped to polynomials of several variables could be interpreted as representations for a parametrized set of space-time possibly representable as space-time surfaces in complexified M^8 (complexified octonions) and mappable to H by $M^8 - H$ duality. They would define a sub-WCW in the "world of classical worlds" (WCW) consisting of space-time surfaces with the same Galois group defined by the roots of X_n. WCW spinor fields would be restricted to this sub-WCW as fermionic Fock states associated with the corresponding space-time surfaces.

   Symmetric polynomials do not correspond to infinite primes but one can wonder whether one could construct WCW spinor fields invariant with respect to S_n or its subgroup. A possible interpretation of S_n in terms of Bose-Einstein statistics and of $G \subset S_n$ in terms of anyonic statistics was already mentioned.

This picture strengthens the hope that TGD might be formulated as a generalized number theory with infinite primes forming the bridge between classical and quantum such that real numbers, p-adic numbers, and various generalizations of p-adics emerge dynamically from algebraic physics as various completions of the algebraic extensions of rational complexified quaternions and complexified octonions. Complete algebraic, topological and dimensional democracy would characterize the theory.

4.4 About possible physical implications

4.4.1 The order of Galois group equals to the dimension of extension

The order of the Galois group is equal to the dimension of the extension. For S_n the order is $n!$ and for the simple group $A_n = S_n/Z_2$, is $n!/2$. The dimension $n! \simeq \sqrt{2\pi n}(n/e)^n$ as the number of space-time sheets increases roughly exponentially with n (https://cutt.ly/Omk8K6E). For instance, the order of A_{34} is $34!/2 \simeq 1.4710^{38}$ whereas Mersenne prime M_{127} is $M_{127}2^{127} - 1 \simeq 1.7 \times 10^{38}$. There might be a correlation between p-adic length scales and dark scales proportional to n reflecting resonant coupling between phases with different values of n when dark scale and p-adic length scale are nearly identical.

The Galois group need not act in CP_2 direction and the orbits of the Galois group can quite well be in M^4 direction. Coherent structures of parallel flux tubes with a very large number of flux tubes are suggestive.

The gravitational Planck constant $h_{eff} = h_{gr} = GMm/v_0$, where v_0 is a parameter with dimensions of velocity, has very large values, and extensions with very large dimensions of extension could be assigned with gravitational flux tube bundles.

Galois group acts on the Fock states of fermions. A natural expectation is that this space is effectively finite-dimensional and can be regarded as the $2^{n/2}$-dimensional space of spinors for an n-dimensional extension.

The hierarchy of infinite primes, which correspond to a hierarchy of polynomials in n variables with a natural action of S_n. n has a logarithmic dependence on the order of the Galois group. The Galois groups representable as sub-groups of S_n are representable also as sub-groups of S_{n+1} so that there is an inclusion hierarchy.

4.4.2 Most groups of order at most n are 2-groups

All p-groups can act as Galois groups of EQ. Most groups of order at most n are 2-groups (https://en.wikipedia.org/wiki/P-group). This result is highly interesting from the TGD point of view.

p-Adic length scale hypothesis $p \simeq 2^k$ if the order $n = 2^k$ for Galois group correlates with p-adic prime $p \simeq 2^k$. There is also support for a more general form of hypothesis for primes near a power of 3 are involved. Could p-adic length scale hypothesis relate directly to the fact that most Galois groups are 2-groups?

The order n of the Galois group corresponds to effective Planck constant $h_{eff} = nh_0$ proportional to the number of sheets of the covering defined by Galois extension. Dark Compton length scales are proportional to n.

I have proposed that dark scales for h_{eff} and p-adic length scales for $h_{eff} = h$ interact in the sense that the transitions between states with $h_{eff} = nh_0$ and $h_{eff} = h = n_0 h_0$ occur resonantly when the p-adic length scale defined for $h_{eff} = h$ is equal to the dark scale [4]. A kind of frequency or wavelength resonance would take place.

If this proposal is correct, the p-adic length scale hypothesis could be understood as a poorly group theoretical fact. A given element of a 2-group has order 2^m, where m depends on the element, and the order 2-group is 2^k. Could these facts have some interpretation in terms of Boolean algebra of 2^k elements? Could group multiplication and collections of elements coming as powers of a generating element have some interpretation in terms of Boolean algebra and provide it with an additional group structure.

Interestingly, the spinor space of the extension would be 2^{2^k}-dimensional if the Galois group is 2-group. This might relate to the proposal that the Combinatorial Hierarchy, which consists of Mersenne primes M(k+1)=$M_{M_k} - 1$, $M^k = 2^k - 1$. One has $M(1) = 3$, $M(2) = 2^3 - 1 = 7$, $M(3) = 2^7 - 1 = 127$, $M(4) = 2^{127} - 1$. It is not known whether the subsequent Mersenne numbers are primes. M^3 is assigned to genetic code and $M(4)$ to possible memetic code. The number of genetic codons is $2^{M_3-1} = 2^6$ and the number of memetic codons would be $2^{M_7-1} = 2^{126}$.

4.4.3 Braids, knots, and Galois groups

The polynomials defining the space-time surface are polynomials of a complex variable. Could one also consider rational functions with n punctures as poles having Taylor or Laurent expansion. This would bring in the braid group B_n as a covering group of S_n and make it possible to find extensions of Q or some EQ in terms of the values of the characters of $G \subset N_n$.

Braid groups also emerge in another way. Partonic 2-surfaces contain fermions as punctures, which suggests that this approach applies to partonic 2-surfaces and that there is a connection with fermionic braids at the light-like orbits of partonic 2-surfaces and that the Galois group represents the braid group.

4.5 Galois groups and genetic code

Abelian groups Z_p, p prime, are simple and the alternating group A_5 with order 60 is the smallest non-Abelian simple group. All groups A_n, $n \geq 5$ are simple and have $n!/2$ elements. A_5 corresponds to the icosahedral group isomorphic with the symmetry group of the dodecahedron.

The TGD based model of genetic code [8, 20, 22] involves in an essential manner the groups A_5 (icosahedron) and A_4 (tetrahedron). Simple groups play a fundamental role in the TGD view of cognition. Could this mean that genetic code represents the lowest level of an infinite cognitive hierarchy?

4.5.1 The TGD inspired model model of genetic code, cognition, and Galois groups

TGD based model of bioharmony [8, 20, 22] provides a model of genetic code as a fusion of 3 icosahedral Hamiltonian cycles and the unique tetrahedral Hamiltonian cycle (what "fusion" precisely means is far from clear and I have considered several options).

Icosahedral Hamiltonian cycles is a non-self-intersecting path at icosahedron connecting nearest points if icosahedron going through all 12 points of the icosahedron. It is interpreted as a representation of a 12-note scale with a scaling by quint assigned to a given step along the cycle. For a given Hamiltonian

ISSN: 2153-8301   Prespacetime Journal   www.prespacetime.com
Published by QuantumDream, Inc.

cycle, the allowed 3-chords of icosahedral harmony are identified as chords defined by the triangular faces of the icosahedron.

Remark: In the sequel I will use the shorthands IH, OH, and TH for icosahedral, octahedral, and tetrahedral harmonies. Also the notation $I/O/T$ will be used for icosahedron/octahedron/tetrahedron unless there is a danger of confusing them with their symmetry groups with identical shorthand notations.

Galois groups are essential for cognition in the TGD framework. In particular, simple groups as primes for groups are also primes for cognition [21]. Genes represent information and Galois groups are crucial for cognition in the TGD framework. Genes would correspond to sequences of 3-chords of bioharmony. This raises several questions.

Could genetic code relate to Galois group A_5 as the smallest simple non-abelian Galois group (and also to the fact that the only polynomials of order smaller than 5 are generically solvable)? Could genetic code correspond to the lowest level in a hierarchy of cognition and of analogs of genetic code?

The order $n = 60$ for A_5 suggests a fusion of 3 icosahedral codes to give $20+20+20 = 60$ codons.

1. 3 Platonic solids, - icosahedron (I), tetrahedron (T) , and octahedron (O) - which have triangles as faces so that one can consider the possibility of constructing a lattice like structure by gluing these Platonic solids together along their faces. Hyperbolic space H^3 indeed allows isosa-tetrahedral tessellation, which also involves O:s. I have proposed that this allows a realization of genetic code and also of genes [22]. The notion of gene generalizes so that genes can also be 2- or 3-D lattice-like structures.

2. A_5 has $A_3 = Z_3$ as a subgroup and I(cosahedron) corresponds to A_5/Z_3. I has several Hamiltonian cycles having as a symmetry group Z_6, Z_4 or Z_2. Z_2 can act either as rotations or reflections.

   Q: Could A_5 as a Galois group as 3-fold covering of I make it possible to understand why the fusion of just 3 icosahedral codes is possible?

3. Tetrahedral group T corresponds to the alternating group $A_4 = S_4/Z_2 = Z_4 \times Z_3$ with 12 elements and tetrahedron identification as A_4/Z_3. The tetrahedral Hamiltonian cycle (4-scale) is unique and has 4 3-chords. The 3-fold copy would correspond to A_4. Information about the unique Hamiltonian cycles of O and T can be found in [1] (`https://cutt.ly/9mlMiV8`).

   Q: Could the factor that there is only one tetrahedral cycle explain why only a single tetrahedron contributes?

4. Octahedral group O has 24 elements and is the wreath product of Z_3 and Z_2^3 and has also the decomposition $O = S_2 \times S_4$. Octahedron can be identified as O/Z_3. Also octahedral Hamiltonian cycle representing 8-scale with 8 chords is unique.

   Q: Why don't octahedral codons contribute?

4.5.2 A model of the genetic code based on icosa-tetrahedral tessellation of hyperbolic 3-space

TGD leads to a proposal for a geometric representation of the genetic code in terms of icosa-tetrahedral tessellation of the hyperbolic 3-space H^3 (mass shell or light-cone proper time $a = constant$ hyperboloids of M^4) [22]. Both I, O, and T having triangular faces appear in the tessellation. Recall that the corresponding harmonies are denoted by IH, OH and TH.

I do not completely understand the details of the icosa-tetrahedral tessellation. The following picture satisfies the constraints coming from the notion of harmony but I have not proven that it is correct. Here the help of a professional geometrician knowing about tessellations of H^3 would be needed.

1. The analog of the discrete translational symmetry for lattices can be assumed: all I:s , O:s and T:s are equivalent as far as common faces with neighboring Platonic solids are considered.

2. The term icosa-tetrahedral tessellation suggests that all octahedral faces are glued to tetrahedral and icosahedral faces so that octahedral chords reduce to either icosahedral or tetrahedral chords. OH would not be an independent harmony. This requires that the number of common faces between two O:s vanishes: $n_O^O = 0$.

3. T shares at least 1 face with a given I so that the number of tetrahedral chords is reduced to at most 3 for given T. 4 purely tetrahedral faces (not shared with I) are needed. I would have $n_{IT} \leq 4$ purely tetrahedral faces in such a way that the total number of purely tetrahedral 3-chords is 4.

   The simplest possibility is that I shares a common face with 2 T:s. Each T shares 2 faces with O providing 2 purely tetrahedral 3-chords and shares the remaining 2 faces with distinct I:s. One would have $n_T^I = 2$, $n_T^O = 2$, $n_T^T = 0$.

   Since each I defines independently 20 chords, 2 I:s cannot have common faces. One would have $n_I^T = 2$, $n_I^I = 0$ and $n_I^O = 18$ to give $n_I^T + n_I^O + n_I^I = 2 + 18 + 0 = 20$.

4. What remains to be fixed are the numbers n_O^I and n_O^T satisfying $n_O^I + n_O^T = 8$. The conditions $n_O^T \geq 1$ and $n_O^I \geq 1$ must be satisfied since both T and I share faces with Os.

   Music comes to rescue here. The 8 3-chords of OH could define OH sub-harmony of IH. Analogously, the 4 3-chords of TH could define TH as a sub-harmony of OH.

   Could IH sharing 18 3-chords with OH contain 2 transposed copies of OH plus 2 chords of TH? IH cannot of course contain the entire TH as a sub-harmony.

   Could OH contain one copy of TH? This would give $n_O^I = n_O^T = 4$. Could the IH part of OH actually be TH as a sub-harmony of IH so that OH would reduce to 2 copies of TH?

To sum up, if the answers to the questions are positive, the incidence matrix n_i^j, $i, j \in \{I, T, O\}$, telling how many faces i shares with j would be given by

$$\begin{bmatrix} n_I^I & n_I^O & n_I^T \\ n_O^I & N_O^O & n_O^T \\ n_T^I & N_T^O & n_T^T \end{bmatrix} = \begin{bmatrix} 0 & 18 & 2 \\ 4 & 0 & 4 \\ 2 & 2 & 0 \end{bmatrix} . \tag{4.1}$$

4.5.3   3-fold cover of the icosa-tetrahedral tessellation

The proposed model does not yet explain the fusion of 3 icosahedral Hamiltomnian cycles. A 3-fold cover of the icosa-tetrahedral tessellation which replaces Platonic solids with their symmetry groups is highly suggestive. This raises a series of questions.

1. How could this representation relate to a possible interpretation in terms of the Galois groups $I = A_5$ and $O = S_2 \times S_4$ and $T = A_4$? Z_3 appears as a sub-group of all these groups and these Platonic solids are coset spaces I/Z_3, O/Z_3, and T/Z_3.

2. Could one lift the icosa-tetrahedral tessellation to a 3-sheeted structure formed by the geometric representations of the Galois groups of this structure acting as symmetry groups? Platonic solids would be replaced with their symmetry groups acting as Galois groups.

3. Could the 3 different icosahedral Hamiltonian cycles correspond to different space-time sheets - roughly CP_2 coordinates as 3-valued functions of M^4 coordinates whereas 20 regions representing icosahedral vertices would correspond to different loci of $E^3 \subset M^4$ just as one intuitively expects?

4. Same should apply to the tetrahedral and octahedral parts of the tessellation. But don't the 3 identical copies of the tetrahedral Hamiltonian cycle give 64+8=72 codons? How can one overcome this problem?

ISSN: 2153-8301   Prespacetime Journal   www.prespacetime.com
                              Published by QuantumDream, Inc.

The following is a possible answer to these questions.

1. $h_{eff} = 60h_0$ corresponds to 60-sheeted space-time (here also $60k$-sheeted space-time is possible if 60-D extension of k-dimensional extension is in question). For T and O an analogous picture would apply. One could say that the projections of I and O and T are in M^4. At each sheet one would have icosa-tetrahedral tessellation.

2. I has 3 types of Hamiltonian cycles with symmetry groups Z_6, Z_4, and Z_2 and can give 3 different copies. However, only a single copy of tetrahedral harmony appears in the model: otherwise the number of codons would be larger than 64. Could the 3 identical Hamiltonian cycles for T and O effectively correspond to a single Hamiltonian cycle?

3. The fusion of Hamiltonian cycles is analogous to a formation of many-boson states. For T and O all Hamiltonian cycles would be identical: one would have only one Hamiltonian cycle effectively. The 3-chords associated with the 3 octahedral and tetrahedral cycles are identical so that only single tetrahedral harmony would be present.

To sum up, the lift of the icosa-tetrahedral complex to that defined by the respective Galois groups could explain why just 3 icosahedral Hamiltomian cycles and effectively only 1 tetrahedral cycle.

Received August 4, 2021; Accepted December 4, 2021

References

[1] Sequin CH. Hamiltonian Cycles on Symmetrical Graphs, 2015. Available at: `https://cutt.ly/EmQFtF8`.

[2] Zeeman EC. *Catastrophe Theory*. Addison-Wessley Publishing Company, 1977.

[3] Rajnbar S Ranjbar F. Inverse Galois Problem and Significant Methods, 2015. Available at: `https://arxiv.org/pdf/1512.08708.pdf`.

[4] Pitkänen M. Magnetic Sensory Canvas Hypothesis. In *TGD and EEG*. Available at: `http://tgdtheory.fi/pdfpool/mec.pdf`, 2006.

[5] Pitkänen M. Massless states and particle massivation. In *p-Adic Physics*. Available at: `http://tgdtheory.fi/pdfpool/mless.pdf`, 2006.

[6] Pitkänen M. Recent View about Kähler Geometry and Spin Structure of WCW . In *Quantum Physics as Infinite-Dimensional Geometry*. Available at: `http://tgdtheory.fi/pdfpool/wcwnew.pdf`, 2014.

[7] Pitkänen M. TGD as a Generalized Number Theory: Infinite Primes. In *TGD as a Generalized Number Theory: Part I*. Available at: `http://tgdtheory.fi/pdfpool/visionc.pdf`, 2019.

[8] Pitkänen M. Geometric theory of harmony. Available at: `http://tgdtheory.fi/public_html/articles/harmonytheory.pdf.`, 2014.

[9] Pitkänen M. Does $M^8 - H$ duality reduce classical TGD to octonionic algebraic geometry?: part I. Available at: `http://tgdtheory.fi/public_html/articles/ratpoints1.pdf.`, 2017.

[10] Pitkänen M. Does $M^8 - H$ duality reduce classical TGD to octonionic algebraic geometry?: part II. Available at: `http://tgdtheory.fi/public_html/articles/ratpoints2.pdf.`, 2017.

[11] Pitkänen M. Does $M^8 - H$ duality reduce classical TGD to octonionic algebraic geometry?: part III. Available at: `http://tgdtheory.fi/public_html/articles/ratpoints3.pdf.`, 2017.

[12] Pitkänen M. p-Adicization and adelic physics. Available at: `http://tgdtheory.fi/public_html/articles/adelicphysics.pdf.`, 2017.

[13] Pitkänen M. Philosophy of Adelic Physics. In *Trends and Mathematical Methods in Interdisciplinary Mathematical Sciences*, pages 241–319. Springer.Available at: `https://link.springer.com/chapter/10.1007/978-3-319-55612-3_11`, 2017.

[14] Pitkänen M. Philosophy of Adelic Physics. Available at: `http://tgdtheory.fi/public_html/articles/adelephysics.pdf.`, 2017.

[15] Pitkänen M. Some comments related to Zero Energy Ontology (ZEO). Available at: `http://tgdtheory.fi/public_html/articles/zeoquestions.pdf.`, 2019.

[16] Pitkänen M. A critical re-examination of $M^8 - H$ duality hypothesis: part I. Available at: `http://tgdtheory.fi/public_html/articles/M8H1.pdf.`, 2020.

[17] Pitkänen M. A critical re-examination of $M^8 - H$ duality hypothesis: part II. Available at: `http://tgdtheory.fi/public_html/articles/M8H2.pdf.`, 2020.

[18] Pitkänen M. Could quantum randomness have something to do with classical chaos? Available at: `http://tgdtheory.fi/public_html/articles/chaostgd.pdf.`, 2020.

[19] Pitkänen M. Generalization of Fermat's last theorem and TGD. Available at: `http://tgdtheory.fi/public_html/articles/FermatTGD.pdf.`, 2020.

[20] Pitkänen M. How to compose beautiful music of light in bio-harmony? `https://tgdtheory.fi/public_html/articles/bioharmony2020.pdf.`, 2020.

[21] Pitkänen M. The dynamics of SSFRs as quantum measurement cascades in the group algebra of Galois group. Available at: `http://tgdtheory.fi/public_html/articles/SSFRGalois.pdf.`, 2020.

[22] Pitkänen M. Is genetic code part of fundamental physics in TGD framework? Available at: `https://tgdtheory.fi/public_html/articles/TIH.pdf.`, 2021.

[23] Pitkänen M. Is $M^8 - H$ duality consistent with Fourier analysis at the level of $M^4 \times CP_2$? `https://tgdtheory.fi/public_html/articles/M8Hperiodic.pdf.`, 2021.

[24] Pitkänen M. Negentropy Maximization Principle and Second Law. Available at: `https://tgdtheory.fi/public_html/articles/nmpsecondlaw.pdf.`, 2021.

[25] Pitkänen M. Some questions concerning zero energy ontology. `https://tgdtheory.fi/public_html/articles/zeonew.pdf.`, 2021.

[26] Pitkänen M. The idea of Connes about inherent time evolution of certain algebraic structures from TGD point of view. `https://tgdtheory.fi/public_html/articles/ConnesTGD.pdf.`, 2021.

[27] Pitkänen M. Time reversal and the anomalies of rotating magnetic systems. Available at: `https://tgdtheory.fi/public_html/articles/freereverse.pdf.`, 2021.

[28] Pitkänen M. What could 2-D minimal surfaces teach about TGD? `https://tgdtheory.fi/public_html/articles/minimal.pdf.`, 2021.

[29] Pitkänen M and Rastmanesh R. Homeostasis as self-organized quantum criticality. Available at: `http://tgdtheory.fi/public_html/articles/SP.pdf.`, 2020.

[30] Pitkänen M and Rastmanesh R. New Physics View about Language: part I. Available at: `http://tgdtheory.fi/public_html/articles/languageTGD1.pdf.`, 2020.

ISSN: 2153-8301   Prespacetime Journal   www.prespacetime.com
Published by QuantumDream, Inc.

[31] Pitkänen M and Rastmanesh R. New Physics View about Language: part II. Available at: `http://tgdtheory.fi/public_html/articles/languageTGD2.pdf.`, 2020.

[32] Pitkänen M and Rastmanesh R. The based view about dark matter at the level of molecular biology. Available at: `http://tgdtheory.fi/public_html/articles/darkchemi.pdf.`, 2020.

[33] Pitkänen M and Rastmanesh R. Why the outcome of an event would be more predictable if it is known to occur? `https://tgdtheory.fi/public_html/articles/scavhunt.pdf.`, 2021.

ISSN: 2153-8301   Prespacetime Journal   www.prespacetime.com
Published by QuantumDream, Inc.

Article

Comparison between the Berry Phase Model of Superconductivity & the TGD Based Model

Matti Pitkänen [1]

Abstract

Hiroyasu Koizumi has proposed a new theory of superconductivity (SC) based on the notion of Berry phase related with an effective magnetic field assignable to adiabatically evolving systems. The model shares similarities with the TGD inspired view about SC. The article also mentioned anomalies that were new to me. This motivated a fresh look in the TGD inspired model. The outcome was an integration of two separate ideas about supraphases.

1. Space-time surfaces as preferred extremals with CP_2 projection of dimension $D = 2$ or $D = 3$ would naturally correspond to 4-D generalizations of so called Beltrami flows, which are integrable flows defined by the flow lines of the induced Kähler field. The existence of a global coordinate z varying along flow lines requires the integrability of the flow. Classical dissipation is absent so that these surfaces are excellent candidates for the space-time correlates of supra flows. The exponential of z gives a phase factor associated with the complex order parameter of a coherent state of Cooper pairs as a counterpart of the Berry phase. Kähler magnetic monopole flux defines the TGD counterpart of "novel" magnetic field.

2. The identification of supra phases as dark matter as $h_{eff} > h$ phases at magnetic flux quanta (tubes and sheets) implies that Cooper pairs correspond to dark fermions associated with the members of flux tube pair, which actually combine to form a closed flux tube. Also single electrons can define supraflow.

3. The Cooper pairs must be created by bosonic oscillator operators constructed from fermionic oscillator operators by bosonization. This is possible only in 1+1-dimensional situations. Thanks to the Beltrami flow the situation is effectively 1+1-dimensional. Bosonization makes it possible to identify SU(2) Kac-Moody algebra, which has an interpretation in the TGD framework.

The assumption that Cooper pairs reside at the magnetic flux quanta solves the 4 problems of standard framework mentioned by Koizumi: high-Tc SCs have two transition temperatures; electron mass m_e instead of its effective mass m_e^* appears in Thomson moment; the reversible phase transition in an external magnetic field inducing a splitting of Cooper pairs does not involve dissipation; why the erratic calculation of the Josephson frequencies in standard model neglecting the chemical potentials gives a correct result? The formation of the Cooper pairs appears as a condition stabilizing the space-time sheets carrying dark matter and all preferred extremals could satisfy the conditions guaranteeing integrable flow and existence of a phase factor varying along flow lines. Could supra phases exist in all scales? Could the breaking of supra phases be only due to the finite size of the space-time sheets? Could even hydrodynamic flow involve super-fluidity of some kind - perhaps based on neutrino Cooper pairs as speculated earlier?

1 Introduction

Hiroyasu Koizumi has proposed a new theory of superconductivity (SC) based on the notion of Berry phase related with an effective magnetic field assignable to adiababically evolving systems. I learned about the theory from a popular article published in Scitechdaily (`https://cutt.ly/LmS4tO1`).

A more technical description of the model can be found in an article [12])(`https://cutt.ly/WmBkIsp`) by Koizumi. The article has title "Superconductivity by Berry connection from many-body wave functions: revisit to Andreev-Saint-James reflection and Josephson effect".

---

[1] Correspondence: Matti Pitkänen `http://tgdtheory.fi/`. Address: Rinnekatu 2-4 8A, 03620, Karkkila, Finland. Email: matpitka6@gmail.com.

1.1 Summary of Berry phase model

The Berry phase model (BPM) explains SC as an implication of a collective phase for which the Berry phase would be a prerequisite. Berry connection acting on the space of quantum states rather than in the space of gauge fields. My interpretation about the basic aspects of the Berry phase theory formed on basis of abstract of [12] is following:

1. In standard model of super-conductity Cooper pairs form a coherent states which is not an eigenstate of electron number. In the new theory fermion number is conserved for Cooper pairs in collective phase and electrons in single electron phase.

2. If Berry connection is non-trivial, it gives rise to a collective mode that generates supercurrent. This collective mode creates number-changing operators for particles participating in this mode, and these number-changing operators stabilize the superconducting state by exploiting the Cooper instability.

   In the new theory, the role of the electron-pairing is to stabilize the nontrivial Berry connection; it is not the cause of SC: also ordinary electrons in collective phase flow without dissipation.

3. In BCS SCs the simultaneous appearance of the nontrivial Berry connection and the electron-pairing occurs. Therefore, the electron-pairing amplitude can be used as an order parameter for the superconducting state and corresponds to Berry phase. In high-Tc SCs the temperatures for the formation of Cooper pairs and for the appearance of SC are different.

4. Andreev-Saint-James reflection [8] and Josephson effect are explained as consequences of the presence of the Berry connection. Bogoliubov quasiparticles are created by superpositions of creation and annihilation operators and utilized also in the BCS model as a convenient tool to diagonalize the kinetic part of the Hamiltonian. In Berry phase modle they are replaced by particle-number conserving Bogoliubov excitations that describe the transfer of electrons between the collective and single particle modes.

The assumption of the model for Josephson effect inducing critics is that the the current in the junctions consists of electrons rather than Cooper pairs.

1. The model treats Josephson junction as an insulator rather than piece of a super-conductor. The model predicts two distinct cases corresponding depending on whether junction is a) thin or b) thick. For a) the Bogoliubov excitations for the two SCs are assumed to be identical. For b) they are not identified. For a) the effect is a first order effect and for b) a second order effect.

2. In BPM a) explains the AC Josephson effect as first order effect when chemical potential difference is taken into account. The supercurrent would be a flow of electrons brought about by the non-trivial Berry connection, which provides an additional U(1) gauge field besides the electromagnetic one.

   This conclusion is due to the presence of chemical potential difference equal to Coulomb energy in equilibrium, otherwise the Josephson frequency spectrum would come as even integer multiples of Josephson frequency.

3. Case b) is the one considered in the standard theory. The effect is second order effect also in the BCS model. If the chemical potential difference between the two SCs is neglected, the model gives BCS prediction for Josephson frequencies.

   If the the chemical potential difference is taken into account in BCS model, the Josephson frequency spectrum would come as half integer multiples of Josephson frequency. The same prediction follows for b) also in the BPM. Some evidence for this kind of half-odd integer spectrum has been repoorted.

Berry phase theory is highly interesting from TGD view point and the comparison with TGD based view about SC is well-motivated.

1. In TGD space-times can be regarded as surfaces in $M^4 \times CP_2$. The effective magnetic field related to the Berry phase has as its TGD counterpart the monopole flux part of the ordinary magnetic field made possible by the non-trivial homology of CP_2 and having no Maxwellian counterpart. Monopole flux assignable magnetic flux tubes carrying also dark matter as $h_{eff} = n_0 > h$ phases of ordinary matter.

2. The existence of Berry phase corresponds to an existence of a phase defined by an angle like coordinate varying along flow lines of an integrable flow associated with induced Kähler field. Integrability is the geometric condition for the existence of the flow and thus superfluid flow or supracurrent.

   The integrable flow is a 4-D generalization [19, 20, 16] of the notion of 3-D Beltrami (magnetic) field [1, 5, 3, 4]. There is no classical counterpart of dissipation as quantum-classical correspondence suggests and the surfaces in question are minimal surfaces as one expects [45]. Generalized Beltrami flows are possible if the dimension D of CP_2 projection of the space-time surface satisfies $D \leq 4$.

3. The members of Cooper pairs in TGD picture would be associated with parallel flux tubes, which form a closed flux tube in a long enough scale. This kind of connections by flux tube pairs can be formed by reconnection of U-shaped flux tube tentacles between two systems and play crucial role in the TGD based model of quantum biology [47].

4. The formation of Cooper pairs accurs at the level of ordinary matter and the liberation of binding energy in their formation allows their transfer to the flux tubes, where dissipation is absent or at least slower by the large value of h_{eff}. Ordinary macroscopic SCy requires high enough value of h_{eff} making possible long enough U-shaped flux loops and thus flux tube pairs.

   As in BPM, the creation of Cooper pairs stabilizes the flux tubes and makes possible non-dissipative currents of both electrons and their Cooper pairs so that SC as non-dissipative current flow is possible also for electrons.

5. Cooper pairs are in a coherent state at the flux tubes and this gives very simple effective Hamiltonian describing the interactions with ordinary matter. The outcome has a lot of common with the standard theory of SC. Both Cooper pairs and electrons are possible supracurrent carriers. The assumption Cooper pairs are at magnetic flux tubes allows however to circumvent the anomalies of the standard models.

6. The Cooper pairs must be created by bosonic oscillator operators constructed from fermionic oscillator operators by bosonization. This is possible only in 1+1-dimensional situations. Thanks to the Beltrami flow the situation is effectively 1+1-dimensional. Bosonization makes it possible to identify SU(2) Kac-Moody algebra, which has an interpretation in the TGD framework.

1.2 BPM, TGD based model, and the anomalies

The BPM is claimed to solve several basic problems of the standard model of SC. Also TGD based model suggests a solution to these anomalies based on the assumption that electrons and Cooper pairs are dark in TGD sense and reside at magnetic flux tubes.

1. High-Tc-superconductivity remains poorly understood. My understanding of the BPM is too limited to allow how it could increase the understanding in this respect.

2. The presence of 2 transition temperatures means that Cooper pairs emerge at higher critical temperature and SC at a lower critical temperature remains poorly understood. BPM predicts that the presence of Cooper pairs stabilizes the collective phase and is a prerequisite for the Berry phase in turn making possible non-dissipative flow. To me this would suggest that these two transition temperatures are identical.

ISSN: 2153-8301   Prespacetime Journal   www.prespacetime.com
Published by QuantumDream, Inc.

TGD: The flux tube pairs forming closed flux tubes serve as carriers of Cooper pairs. The first transition temperature would give rise to rather short flux tubes and SC in short scales and second transition temperature to rather long flux tube and SC in long scales with a larger value of h_{eff} (the scale of quantum coherence scales like h_{eff}).

3. The experimental finding is that London magnetic moment depends on the real mass m_e of electron rather than effective mass m_e^* . If supracurrent flows at the level of ordinary matter, one would expect the appearance of m_e^*. BPM explains this if the collective phase is separate from the ordinary phase.

   TGD: The dark electrons at magnetic flux tubes would not interact directly with condensed matter so that the real mass would appear in the expression of London moment caused by rotation of dark electrons.

4. It has found that the phase transition from SC to ordinary phase in an external magnetic field does not cause dissipation although one would expect this if Cooper pairs split to ordinary electrons. This can be understood in BPM if electrons in collective phase do not dissipate and thus do not interact with ordinary matter.

   TGD: The dissipation would be absent if the electrons from the split Cooper pairs do not dissipate. This is indeed true. One could thus talk about analogs of supra currents for electrons.

The proposed view encourages several questions. The formation of the Cooper pairs appears as a condition stabilizing the space-time sheets carrying dark matter and all preferred extremals could satisfy the conditions guaranteeing integrable flow and existence of a phase factor varying along flow lines. Could supra phases as non-dissipative phases of also fermionic states exist in all scales? Could the breaking of supra phase property be only due to the finite size of the space-time sheets? Could even hydrodynamic flow involve super-fluidity of some kind - perhaps based on neutrinos or neutrino Cooper pairs as speculated earlier?

2 TGD based model of superconductivity

TGD inspired model of super-conducitivity has developed slowly during years [19, 20] [31][35].

2.1 Brief summary of TGD based model

The breakthrough came around 2005 with the emergence of the idea about a hierarchy of phases of ordinary matter having non-standard value $h_{eff} = nh_0$ of Planck constant having arbitrarly large values and behaving in many respects like dark matter. Super-conducting phases would reside at the magnetic flux tubes carrying monopole flux and large value of h_{eff} would be crucial for their stability even at high temperatures.

2.1.1 General mechanism of superconductivity in TGD framework

The ideas about high temperature SC have evolved gradually as a reaction to experimental input and evelution in the understanding of TGD.

1. The many-sheeted space-time concept suggests a very general mechanism of SC based on a transfer of charged particles from atomic space-time sheets to larger space-time sheets. Later these space-time sheets were identified as magnetic flux tubes carrying as $h_{eff} = nh_0$ phases behaving like dark matter.

   The first guess was that larger space-time sheets are very dry, cool and silent so that the necessary conditions for the formation of high T_c macroscopic quantum phases are met. The criticism against this model was that particles touch all space-time sheets having non-empty Minkowski space projection to the region where the particle is so that thermal equilibrium is generated. Darkness as $h_{eff} > h$ property would allow even same temperature since various energy scales would typically scale like h_{eff} implying thermal stability.

One must however take the assumption about thermal equilibrium with a grain of salt. The TGD based model for the aging of a living system [48] assumes that the space-time sheets carrying dark matter slowly approach thermal equilibrium with the space-time sheets carrying ordinary matter [46]. The slow approach to thermal equilibrium would be due to a small amount of dissipation at flux tubes.

2. The possibility of large h_{eff} quantum coherent phases makes the assumption about thermal isolation between space-time sheets un-necessary. In the model to be discussed in this article Cooper pairs are created at the level of ordinary matter by standard mechanisms and transferred to flux tubes.

3. It became clear quantum criticality predicting a new kind of SC explaining the strange features of high T_c SC is essential. Two kinds of Cooper pairs, or rather flux tubes are assumed. They correspond to a different values of $h_{eff} > h$. Either the Cooper pairs or flux tubes with smaller value of h_{eff} have shorter life time (proportional to h_{eff}). Both Cooper pairs and flux tubes correspond to super-conductivity but in different time and length scales. In the transition to SC in long scales the closed but short flux tubes looking like flux tube pairs reconnect to long flux tubes.

Below temperature $T_{c_1} > T_c$ only the Cooper pairs with smaller value of h_{eff} are present and their short lifetime implies that SC is broken to ordinary conductivity in longer scales satisfying scaling laws characteristic for criticality. At T_c Cooper pairs and flux tubes with longer lifetime become possible and have considerably longer life time.

These two superconducting phases compete in a certain narrow interval around critical temperature T_c for which body temperature of endotherms is a good candidate in the case of living matter.

4. Magnetic flux tubes would be carriers of dark particles and according to the findings about high temperature SC magnetic fields would be crucial for SC. Two parallel flux tubes carrying magnetic fluxes in opposite directions is the simplest candidate for a super-conducting system. This conforms with the observation that antiferromagnetism is somehow crucial for high temperature SC. The spin interaction energy is proportional to h_{eff} and can be above thermal energy: if the hypothesis that dark cyclotron energy spectrum is universal is accepted, then the energies would be in bio-photon range and high temperature SC is obtained. If fluxes are parallel spin $S = 1$ Cooper pairs are stable. $L = 2$ states are in question since the members of the pair are at different flux tubes. These two kinds of Cooper pairs could correspond to BCS type and exotic Cooper pairs.

The fact that the critical magnetic fields can be very weak or large values of $\hbar$ is in accordance with the idea that various almost topological quantum numbers characterizing induced magnetic fields provide a storage mechanism of bio-information.

5. This mechanism of high temperature SC is extremely general and in principle works for electrons, protons, bosonic ions and Cooper pairs of fermionic ions, charged molecules and even neutrinos and an entire zoo of high T_c bio-SCs, super-fluids and Bose-Einstein condensates is predicted. The variant of the model to be discussed in this article predicts that also charged fermionic states give rise to non-dissipative currents and that the formation of Cooper pairs a prerequisite for the $h_{eff} > h$ phase.

6. For gravitational flux tubes the generalization of Nottale hypothesis [14] states that $\hbar_{eff} = \hbar_{gr} = GMm/v_0$ is very large and to the particle mass. Therefore the binding energy of Cooper pairs identifiable as spin-spin interaction energy and does not depend on the mass of the Cooper pair. Supraphases would be universal in this case. This form of superconductivity is proposed to be crucial for living matter.

2.1.2 Quantitative model of high-T_c SC and bio-SC

I have developed already earlier [17, 18, 19, 20] a rough model for high T_c super conductivity [9, 10, 11, 7, 6, 13]. The members of Cooper pairs are assigned with parallel flux tubes carrying fluxes which have either same or opposite directions. The essential element of the model is hierarchy of Planck constants defining a hierarchy of dark matters.

1. In the case of ordinary high T_c SC bound states of charge carriers at parallel short flux tubes become stable as spin-spin interaction energy becomes higher than thermal energy.

   The transition to SC is known to occur in two steps: as if two competing mechanisms were at work. A possible interpretation is that at higher critical temperature Cooper pairs become stable but that the flux tubes are stable only below rather short scale: perhaps because the spin-flux interaction energy for current carriers is below thermal energy. At the lower critical temperature the stability would is achieved and supra-currents can flow in long length scales.

2. The phase transition to SC is analogous to a percolation process in which flux tube pairs fuse by a reconnection to form longer super-conducting pairs at the lower critical temperature. This requires that flux tubes carry anti-parallel fluxes: this is in accordance with the anti-ferro-magnetic character of high T_c super conductivity. The stability of flux tubes very probably correlates with the stability of Cooper pairs: coherence length could dictate the typical length of the flux tube.

3. A non-standard value of h_{eff} for the current carrying magnetic flux tubes is necessary since otherwise the interaction energy of spin with the magnetic field associated with the flux tube is much below the thermal energy.

There are two energies involved.

1. The spin-spin-interaction energy should give rise to the formation of Cooper pairs with members at parallel flux tubes at higher critical temperature. Both spin triplet and spin singlet pairs are possible and also their mixture is possible.

2. The interaction energy of spins with magnetic fluxes, which can be parallel or antiparallel contributes also to the gap energy of Cooper pair and gives rise to mixing of spin singlet and spin triplet. In TGD based model of quantum biology antiparallel fluxes are of special importance since U-shaped flux tubes serve as kind of tentacles allow magnetic bodies form pairs of antiparallel flux tubes connecting them and carrying supra-currents. The possibility of parallel fluxes suggests that also ferro-magnetic systems could allow SC.

   One can wonder whether the interaction of spins with magnetic field of flux tube could give rise to a dark magnetization and generate analogs of spin currents known to be coherent in long length scales and used for this reason in spintronics (`http://tinyurl.com/5cu3qh`). One can also ask whether the spin current carrying flux tubes could become stable at the lower critical temperature and make SC possible via the formation of Cooper pairs. This option does not seem to be realistic.

2.2 TGD counterparts of the collective phase, novel magnetic field, and Berry's phase

In the standard model of superconductivity SC is characterized by a complex order parameter for which the Berry phase would serves as an analog in BPM. Berry phase is a consequence of adiabaticity and characterizes collective phase. One can assign to the Berry phase effective $U(1)$ gauge field which reduces to magnetic field in a static situation. What are the TGD counterparts of these notions?

2.2.1   Beltrami flow as space-time correlate for non-dissipative flow

TGD provides the geometrization of classical physics in terms of space-time surfaces carrying gravitational and standard model field as induced fields so that both the supra current and the phase should have geometric intepretation. This serves as a powerful constraint on the model.

1. Supra current must correspond to a flow. The flow must be integrable in the sense that the coordinate defined along flow lines defines a global coordinate at flux tubes. One can indeed argue that an operational defition of a coordinate system requires that coordinates correspond to coordinates varying along flow lines of some physical flow. The exponential of the coordinate would define the phase factor of the complex order parameter such that its gradient defines the direction of the supracurrent.

   If the motion of particles is random one cannot talk of a hydrodynamic flow but something analogous to the motion of gas particles or Brownian motion. In the TGD framework this situation corresponds to disjoint space-time sheets as a representation of particle orbits. The flow property could however hold true inside the "pieces" of space-time. The coherence scales of flow would become short.

2. One must make it clear that here an approximation is made. Elementary particles have as building bricks wormhole contacts defining light-like partonic orbits to which one can assign light-like curves as M^4 projections. For a vanishing value $\Lambda = 0$ of cosmological constant (real analytic functions at M^8 level), these curves are light-like (light-likeness condition reduces to Virasoro conditions) whereas for $\Lambda > 0$ (real polynomials) at M^8 level the projections consist of pieces which are light-like geodesics somewhat like in the twistor diagrams [45]. Smooth curve is replaced with its approximation.

   For massive particles, this orbit would be analogous to zitterbewegung orbit and the motion in the long scales would occur with velocity $v < c$: this provides a geometric description of particle massiation. The supracurrent would not actually correspond to the flow as such but to CP_2 type extremals along the flow lines.

3. In the Appendix appearing also in [16], I have briefly discussed a decades old proposal that the 4-D generalization of so called Beltrami flow [1, 5, 3, 4], which defines an integrable flow in terms of flow lines of magnetic field, could be central in TGD. Superfluid flows and supra currents could be along flux lines of Beltrami flows defined by the Kähler magnetic field [23, 21].

   If the Beltrami property is universal, one must ask whether even the ordinary hydrodynamics flow could represent Beltrami flow with flow lines interpreted in terms of flow lines Kähler magnetic field appearing as a a part of classical Z^0 field. Could hydrodynamical flow be stabilized by a superfluid made of neutrino Cooper pairs. h_{eff} hierarchy of dark matters in turn inspires the question whether weak length scale could be scaled up to say cellular length scales (neutrino mass corresponds to a length scale of a large neuron).

4. The integrability condition

$$j \wedge dj = 0 \tag{2.1}$$

   of the Beltrami flow states that the flow is of form

$$j = \Psi d\Phi \; , \tag{2.2}$$

   where Φ and Ψ are scalar functions, which means that Ψ defines a global coordinate varying along the flow lines.

5. Beltrami property means that the classical dissipation characterized by the contraction of the Kähler current

$$j^\alpha = D_\beta J^{\alpha\beta} \tag{2.3}$$

with Kähler form $J_{\alpha\beta}$ is absent:

$$j^\beta J_{\alpha\beta} = 0 \ . \tag{2.4}$$

In absence of Kähler electric field (stationary situation), this condition states the 3-D current is parallel with the magnetic field that it creates.

In 4-D case, the orthogonality condition guarantees the vanishing of the covariant divergence of the energy momentum tensor associated with the Kähler form. This condition is automatically true for the volume part of the energy momentum tensor but not for the Kähler part, which is essentially energy momentum tensor for Maxwell's field in the induced metric. As far as energetics is considered, the system would be similar to Maxwell's equations.

The vanishing of the divergence of the energy momentum tensor would support Einstein's equations expected at QFT limit of TGD when many-sheeted space-time is approximated with a slightly curved region of M^4 and gauge and gravitational fields are defined as the sums of correspond induced fields (experienced by test particles touching all space-time sheets).

6. An interesting question is whether Beltrami condition holds true for all preferred extremals [21] [45], which have been conjectured to be minimal surfaces analogous to soap films outside the dynamically generated analogs of frames at which the minimal surface property fails but the divergences of isometry currents for volume term and Kähler action have delta function divergences cancelling each other. The Beltrami conditions would be satisfied for the minimal surfaces.

If the preferred extremals are minimal surfaces and simultaneous extremals of both the volume term and the Kähler action, one expects that they possess a 4-D analog of complex structure [45]: the identification of this structure would be as Hamilton-Jacobi structure [21] to be discussed below.

7. Earlier I have also proposed that preferred extremals involving light-like local direction as direction of the Kähler current and orthogonal local polarization direction. This conforms with the fact that Kähler action is a non-linear generalization of Maxwell action and minimal surface equations generalize massless field equations. Locally the solutions would look like photon like entities.

This inspires the question whether all preferred extremals except CP_2 type extremals defining basic building bricks of space-time surfaces in H have a 2-D or 3-D CP_2 projection and allow interpretation as thickening of flux tubes? CP_2 type extremals have 4-D CP_2 projection and light-like M^4 projection and an induced metric with an Euclidean signature.

2.2.2   Could all conserved currents define integrable flows?

In the TGD framework, the classical field equations for the space-time surface can be regarded as hydrodynamical in the sense that they express the conservation of the currents associated with the isometries of $H = M^4 \times CP_2$ [24]. The classical field equations for the preferred extremals follow as consistency conditions for the modified Dirac equation obeyed by a second quantized induced spinor field [25], whose second quantization is induced by the quantization of free spinor fields of H [26].

An attractive conjecture is that all isometry currents or at least part of them (depending on the situation) are also Beltrami currents. For $j^A = \Psi^A d\Phi^A$ this implies $d\Psi^Q \wedge d\Phi^A = 0$ so that Ψ is function of Φ and j^A is expressible as a gradient of a scalar function: $j^A = d\chi^A$ and defines an integrable flow with a global coordinate varying along the flow lines of the current.

This prediction is obviously very powerful. This condition is linear and could make sense also for the fermionic currents involving bilinears of the oscillator operators. One would have genuine quantum hydrodynamics. In the TGD framework, the classical field equations for the space-time surface can be regarded as hydrodynamical in the sense that they express the conservation of the currents associated with the isometries of $H = M^4 \times CP_2$. The field equations follow as consistency conditions for the modified Dirac equation for a second quantized induced spinor field whose second quantization is induced by the quantization of free spinor fields of H.

The obvious objection against the strong form of the conjecture is that gradient currents are irrotational. This is true in Euclidean space but if the first homology group of the 3-surface is non-trivial. Gradient current can be rotational in the same sense as the vortices of supraflow, which have a quantized circulation concentrated at the axis of rotation.

2.2.3 Some examples

Some special cases help to get some perspective.

1. For $j^\alpha = 0$ condition he condition is trivially true: this is true for CP_2 type extremals. For massless extremals (MEs) the condition is true because of light-likeness of j^α. MEs are proposed to have a generalization with 3-D CP_2 projection.

2. In [23, 16] it is found that for non-trivial solutions the dimension of the CP_2 projection of the space-time surface is $D = 2$ or $D = 3$. $D = 2$ would include string-like objects $X^2 \times Y^2 \subset M^4 \times CP_2$ having a 2-D string world sheet X^2 as an M^4 projection: in this case $j^\alpha = 0$ would hold true so these extremals cannot describe SC. This phase would be highly ordered.

3. $D = 3$ phase would be between order and chaos and extremely complex: in this case j^α could be non-vanishing. The topologies of the flux lines for magnetic fields satisfying the Beltrami condition gives an idea about the complexity. SC would correspond to this situation.

2.2.4 Does the M^4 part of Kähler form produce problems?

The twistor lift of TGD [27, 30] suggests that the Kähler form of H has also M^4 part. M^4 part could give rise to observed small CP breaking and be relevant also for matter antimatter asymmetry [39, 42].

The M^4 Kähler form corresponds to an analog of a self-dual instanton field for which E and B are constant, orthogonal and have the same strength so that the action vanishes for the canonically imbedded M^4. Physically M^2 is characterized by light-like direction and E^2 complex coordinate. This field selects a preferred decomposition $M^2 \times E^2$ of M^4 and breaks Lorentz invariance. How can one save Lorentz invariance?

One can also consider local selections of polarization and light-like momentum directions. I call Hamilton-Jacobi structures [21] and they provide a concrete realization of the analog of complex structure in the case of M^4.

Hamilton-Jacobi structure is an integrable distributions of M^2 and E^2 defining slicings of M^4 by string world sheets having an orthogonal Euclidian 2-surface at each point. The moduli space for the Hamilton-Jacobi structures serves as the analog of the moduli space of complex structures for 2-D surface. Hamilton-Jacobi structure should not be God-given but be dynamically determined and part of WCW.

The existence of $M^8 - H$ duality implies this. The construction of $X^4 \subset M^8$ assigns to it $M^4 \subset M^8$. Space-time surface $X^4 \subset M^8$ as a "root" of an octonionic polynomial associates to M^4 a Hamilton-Jacobi structure. This makes it possible to parametrize the tangent spaces of $X^4 \subset M^8$ by CP_2 coordinates and therefore $M^8 - H$ duality as a map $X^4 \subset M^8 \to X^4 \subset H$.

Could the contribution of M^4 Kähler action to the total Kähler current spoil the minimal surface property by spoiling the analytic structure? Could the existence of a 4-D analog of the complex structure and implying minimal surface property prevent this? Note that Beltrami flow property is not lost if the contribution $j^\alpha(M^4)$ to Kähler current vanishes.

2.3 Coherent states and the problem with fermion number conservation

The number of electrons and Cooper pairs are ill-defined for SC. This is required by the existence of an order parameter ψ having a well-defined phase. In the phase space picture of the harmonic oscillator phase angle is a conjugate of the radial phase space coordinate, whose quantized value in the Bohr model characterized by an integer n characterizing the energy eigenvalues of the harmonic oscillator. In quantum field theory n has interpretation as the number of particles in a given mode. Phase is well-defined for coherent states for Cooper pairs, which are eigenstates of annihilation operators of Cooper pairs. In QFTs the eigenvalues of annihilation operators define analogs of Fourier components of classical fields.

One can argue that the assumption of ill-defined fermion number and energy is unphysical. In the TGD framework one can consider two solutions to the problem.

1. Zero energy ontology (ZEO) provides the first candidate for a solution. In ZEO quantum state is a superposition of deterministic time evolutions and by holography equivalent to a superposition of pairs of ordinary 3-D quantum states located at the boundaries of causal diamond (CD) identified as intersection of future and past directed light-cones. These 3-D states have the same total quantum numbers and for keeping purposes their quantum numbers can be taken to be opposite so that the entire state has zero quantum numbers. Zero energy state can be a superposition of states for which the 3-D states at either boundary with varying quantum numbers such as energy and fermion number. There are no problems with the conservation of fermion number and energy. The density matrix describing the entanglement between the 3-D states at the opposite boundaries of CD is non-trivial for these states and the interpretation in terms of a thermal state is attractive.

2. Second solution is that the system is not closed. The total number of electrons and total energy are well-defined only for the system consisting of ordinary matter and dark matter at magnetic flux tubes. Superconductivity would be direct proof of the reality of dark matter. The transition to super-conductivity would transfer Cooper pairs formed at the level of ordinary matter to the magnetic flux tubes as dark phase.

   The collective phase proposed in BPM is analogous to the dark matter at flux tubes. The novel magnetic field as an effective magnetic field assigned with the Berry phase would correspond in TGD framework to Kähler magnetic field at flux tubes carrying monopole flux not possible in Maxwelian world.

   This option seems to be the realistic one.

2.3.1 Bosonization requires effective 1+1-dimensionality

It is convenient to denote the oscillator operators for electrons at the level of ordinary matter by $b_k^\dagger$ and b_k and oscillator operators for Cooper pairs at flux tubes by $c_m^\dagger$ and c_m. They are assumed to satisfy standard anticommutation/commutation relations.

1. The bosonic oscillator operators c_N creating Cooper pairs must be representable as superpositions of electron pairs. An even stronger condition is that a subset of fermionic oscillator operator pairs are representable as bosonic oscillator operators. This requires what is known as bosonization. Bosonization was discovered independently by particle physicists Sidney Coleman and Stanley Mandelstam and condensed matter physicists Daniel C. Mattis and Alan Luther. Unfortunately the Wikipedia article about bosonization (`https://cutt.ly/HmGYPnM`) is very confusing and it is better to read the article [2] (`https://cutt.ly/BmGNzeA`) about bosonization.

Remarkably, bosonization is possible only when the system is effectively 1+1-dimensional.

2. One considers chiral fermions for which the spinor fields with different helicities are decomposed to parts ψ with wave vectors $k > 0$ and $\overline{\psi}$ with wave vectors $k < 0$.

$$\psi = \int_{k>0} \frac{dk}{2\pi} [exp(ikx)\alpha(k) + exp(-ikx)\beta^\dagger(k)] \ ,$$

$$\overline{\psi} = \int_{k<0} \frac{dk}{2\pi} [exp(ikx)\alpha(k) + exp(-ikx)\beta^\dagger(k)] \ . \tag{2.5}$$

It should be noticed that the definition of $\overline{\psi}_+$ does not involve hermitian conjugation as usually.

3. ψ is expressed in terms of bosonic field ϕ as

$$\psi =: exp(i \int_\infty \partial_{++}\phi) : \ , \quad \overline{\psi} =: exp(-i \int_\infty \partial_{++}\phi) : \ . \tag{2.6}$$

The subscript $\pm$ refers to either light-like coordinate.

The bosonic and fermionic currents are related by

$$\partial_{++}\phi =: \overline{\psi}\psi : \ . \tag{2.7}$$

Note that the right hand side has fermion number 2. The condition $\partial_{+-}\phi = 0$ is satisfied and corresponds to massless d'Alembert equation in 1+1 dimensions.

Coherent state is an eigenstate of ϕ and therefore of $\partial_{++}\phi)$ and thus an eigenstate of the supracurrent.

4. The explicit formulas for the bosonization are given in the book "Field Theories of Condensed Matter Physics" by Eduardo Fradkin [2] in the chapter about Luttinger liquid (page 164). Although this model does not apply as such in the TGD framework, it gives an idea about the construction.

The bosonized expression fermionic oscillator operators with opposite spins and chiralities are mapped to the bosonized variants by the rule

$$\psi^\dagger_{R,\uparrow} \to exp(i\sqrt{2\pi}\theta_c) \ , \quad \psi^\dagger_{R,\downarrow} \to exp(-i\sqrt{2\pi}\theta_c) \ ,$$
$$\psi^\dagger_{L,\uparrow} \to exp(i\sqrt{2\pi}\Phi_s) \ , \quad \psi^\dagger_{L,\downarrow} \to exp(-i\sqrt{2\pi}\Phi_s) \ . \tag{2.8}$$

In the singlet case, these rules give the correspondences

$$O_{SS} = \psi^\dagger_{R\uparrow}\psi^\dagger_{L\downarrow} \ \to \ exp(i\sqrt{2\pi}\theta_c)exp(-i\sqrt{2\pi}\Phi_s) \ . \tag{2.9}$$

In the triplet case, one obtains

$$O^1_{TS} = \psi^\dagger_{R\uparrow}\psi^\dagger_{R\uparrow} \ \to \ exp(i\sqrt{2\pi}\theta_c)exp(i\sqrt{2\pi}\phi_s) \ ,$$
$$O^{-1}_{TS} = \psi^\dagger_{R\downarrow}\psi^\dagger_{R\downarrow} \ \to \ exp(-i\sqrt{2\pi}\theta_c)exp(-i\sqrt{2\pi}\phi_s) \ 0. \tag{2.10}$$

ISSN: 2153-8301   Prespacetime Journal   www.prespacetime.com
Published by QuantumDream, Inc.

One must generalize these formulas to the TGD framework for a given flux tube. It is important to notice that there is quantum superposition over different flux tube configurations in the "world of classical worlds" (WCW) so that the inclusion of WCW degrees of freedom not present in QFT description is unavoidable.

1. The fermionic modes of opposite spin are defined at the same closed flux tube. Whether one should restrict the fields with opposite chirality to different flux tube portions is an open question.

2. The fermionic oscillator operators at space-time surface are labelled by a longitudinal momentum like quantum number, which in suitable units for closed flux tube allowing in a good approximation as a straight flux tube locally becomes integer valued momentum locally parallel to the flow line - the momentum scale is determined by Fermi momentum.

   The members of pairs have momenta $P_\pm = p_{cm}/2 \pm k$, where k has magnitude of order Fermi momentum, and p_{cm} is the total longitudinal momentum, which has an upper bound below Fermi momentum. The transversal quantum numbers are integer valued using as a basic unit $p_{min} = \hbar_{eff}/L$, where L of the order of the length of the flux tube.

   Since one has $p_{cm} = \hbar_{eff}/\lambda$ for Cooper pairs, their wavelengths are scaled up by the ratio $\hbar_{eff}/\hbar$ from their normal values. Also the length L of the flux tube is scaled up in this way from that for $h_{eff} = h$.

3. Additional quantum numbers are angular momentum eigenvalue m in the local flux tube direction and harmonic oscillator quantum number n labelling cyclotron states. The most plausible option is that one has phases characterized by the values of n and n.

   The breaking of rotational symmetry caused by the magnetic field takes place for a given space-time surface in the superposition. For $m > 0$ the angular momentum eigenvalues of cyclotron states contribute and one obtains Cooper pairs with relative angular momentum. Fermi statistics allows only even integer valued total angular momentum.

2.3.2 Kac-Moody symmetry associated with the bosonization

In [2] it is mentioned that the bosonization gives rise to SU(2) Kac-Moody algebra such that I_3 generators is generated by the ϕ and generators $I_\pm$ by normal order exponentials of ϕ. This construction is applied in string models by extending the Cartan algebra represented by scalar fields to the entire algebra.

TGD predicts that the isometries of H give rise to an extended Kac-Moody algebra assignable to the 3-D light-like orbits of the partonic 2-surfaces at which the signature of the induced metric changes from Minkowskian to Euclidian.

This algebra is localized not only with respect to the complex coordinate z of the partonic 2-surface but also with respect to the light-like coordinate r varying along the partonic orbit. The extension is possible because light-likeness implies metric 2-dimensionality.

General coordinate invariance motivates the question whether this Kac-Moody algebra extends to a slicing by light-like 3-surfaces parallel to the partonic orbits.

2.3.3 Bosonization requires Beltrami property

Bosonization requires effective 1+1-dimensionality. This is guaranteed by the Beltrami flow property of supra currents. In TGD all fermionic oscillator operators at space-time surface are representable in terms of oscillator operators associated with the spinor harmonics of $H = M^4 \times CP_2$. The existence of Beltrami flow implies the existence of single preferred coordinate assignable to the flux tubes and if the transversal degrees of freedom are frozen for the Cooper pairs in given phase of SC, the system is effectively 1-D.

One can consider a variety of phases in which the cyclotron excitations assignable to the transversal degrees of freedom assignable are present. These cyclotron states and transitions between them play a key role in TGD inspired view about quantum biology.

ISSN: 2153-8301   Prespacetime Journal   www.prespacetime.com

Published by QuantumDream, Inc.

2.3.4 Why the formation of Cooper pairs is necessary for the formation of $h_{eff} > h$ dark phase?

Why would the formation of Cooper pairs be necessary for the formation of the dark phase? Here the understanding of the energetics $h_{eff} > h$ phases helps.

1. Quite generally, the energy of the quantum state increases with h_{eff} so that the creation of the dark electrons requires energy.

   This energy would be provided in the formation of the Cooper pair as the liberated binding energy. Cooper pairs would be formed already at the level of the ordinary matter. The bosonic field modelling Cooper pairs would couple to 2-electron bilinear characterizing the quantum state of the Cooper pair.

   Since Cooper pairs are formed at the level of ordinary matter, the view of the formation of Cooper pairs is consistent with the conventional picture involving photons and effective attractive interaction generated by the attractive interaction between electrons and atoms.

2. In this process fermion number decreases by 2 units (in the recombination of electron and hole it would decrease by 1 unit). This process is analogous to Andreev-Saint-James reflection [8] (`https://cutt.ly/AmDYDTG`), which could therefore be seen as direct evidence for the transfer of electron pairs to magnetic flux tubes. Andreev-Saint-James reflection occurs at the normal metal-SC interface and gives rise to lower energy states at the surface of unconventional SC.

2.4 The general form of the effective Hamiltonian

Consider now the general form of the effective Hamiltonian H_{eff} obtained from a quartic Hamiltonian in oscillator operators of H spinor field at space-time surface.

1. The effective Hamiltonian operator H_{eff} modelling the system would be formed as a linear in the oscillator operators $c_k^\dagger$ (c_k) creating (annihilating) Cooper pairs at flux tubes and products $b_k b_l$ ($b_k^\dagger b_l^\dagger$) of annihilation (creation) operators for ordinary electrons.

   If also free electrons are possible at flux tubes, H_{eff} contains also a part, which is bilinear both in the electronic oscillator operators at flux tubes and at the level of ordinary matter.

2. H_{eff} contains a term of form $H_1 = H_2 + H_2^\dagger$

$$H_2 = C^{Nkl} c_N b_k^\dagger b_l^\dagger \ . \tag{2.11}$$

3. For coherent states of Cooper pairs the action of the annihilation operators c_N reduces to a multiplication with a complex number C_N so that H_2 reduces to a kinetic term

$$H_2 = B_{kl} b_k^\dagger b_l^\dagger \ , \quad B_{kl} = C^{Nkl} C_N \ . \tag{2.12}$$

   The kinetic part has the same form as the kinetic term of H_{eff} in the standard model for SC. One can diagonalize this part of Hamiltonian by a Bogoliubov transformation `https://cutt.ly/DmDcbC7` mixing the creation and annihilation operators for electrons with different quantum numbers. Bogololiubov transformation can be regarded as a symplectic transformations at the level of phase space.

ISSN: 2153-8301   Prespacetime Journal   www.prespacetime.com
Published by QuantumDream, Inc.

4. The remaining terms can be treated as a perturbation. $H_2^\dagger$ is of form

$$H_2^\dagger = \overline{C}^{Nkl} c_N^\dagger b_k b_l \ \ .$$

(2.13)

$H_2^\dagger$ makes possible the transfer of electron pairs to the flux tubes as Cooper pairs. It also makes possible the Andreev-Saint-James reflection regarded as the reflection of the electron as a hole from the boundary of SC.

5. H_{eff} contains also a quartic term quadratic in electronic oscillator operators both at the flux tubes and at the level of ordinary matter. This term makes possible the transfer of electrons to electron pairs not forming Cooper pairs at flux tubes.

   The oscillator operators $b_k(tube)$ at flux tube creating single fermion states should correspond to oscillator operators not appearing in $\partial \phi_{++} =: \overline{\psi}\psi :$.

6. The assumption that a closed flux tube forming effectively a flux tube pair is involved suggests that the members of the Cooper pair are at different flux tubes. If this is the case the fermionic oscillator operators at different flux tubes anticommute and the commutator of the bosonic oscillator operators does not involve bi-local terms.

2.5   A more precise formulation of TGD based theory by starting from BCS theory

It is instructive to see whether BCS theory could allow a more detailed formulation of the TGD inspired theory. The Wikipedia article (`https://cutt.ly/4mBkA5i`) gives a good summary of BCS theory.

1. Electrons of the lattice are treated as free Fermi gas and at zero temperature electrons are below the Fermi surface. In the simplest situation, Fermi surface is a sphere defined by Fermi energy (`https://cutt.ly/UmBkFhb`)

$$E_F = \frac{p_F^2}{2m_e} = \frac{\hbar^2}{2m_e}(3\pi^2 n_e)^{2/3} \ \ .$$

(2.14)

   Here n_e is the density of conduction electrons. Fermi temperature is equal to Fermi energy in the natural units. Examples of the values of the Fermi energy, Fermi temperature, Fermi velocity, and electron number density can be found in `https://cutt.ly/zmBkHt7`.

2. Fermi statistics implies that the transition to super-conductivity involving formation of Cooper pairs occurs for electrons near the Fermi surface. Any attractive interaction between electrons can cause the creation of Cooper pairs and the mechanism based on the interaction with phonons is the mechanism in BCS theory.

2.5.1   Critical temperature as Hagedorn temperature for magnetic flux tubes

The transition to super-conductivity involves an exponential increase in the heat capacity. This could be seen as a support for the flux tube picture. Flux tubes are string like objects and have an infinite number of degrees of freedom and the feed of energy excites these degrees of freedom so that temperature increases very slowly.

This implies a maximal temperature known as Hagedorn temperature T_H in string model context. The identification $T_c = T_H$ is suggestive. Living matter can remain functional in a rather narrow range of temperatures. I have proposed that the critical temperature corresponds to Hagedorn temperature [46] for the magnetic body of the system receiving information from and controlling the biological body.

If the identification of T_c as Hagedorn temperature for the magnetic flux tubes is correct, the spectrum of the critical temperatures could be universal. On the other hand, in the general BCS model of SC, critical temperature depends on the mechanism for the formation of Cooper pairs.

1. For attractive interaction caused by phonon vibrations one has

$$T_c = 1.134 \times E_D \times exp(-\tfrac{1}{N(0)}) \ , \quad N(0) = n(0)V \ . \tag{2.15}$$

Here $N(0)$ is the total number of conduction electrons at $T = 0$ and $E_D = \hbar\omega_D$ is Debye energy defined as the maximal value of frequency $f_D = c_s/\lambda_D$ for sound wave defined by the minimal wavelength λ_D by the minimal size of objects involved in the oscillations. The size of the lattice cell gives an order of magnitude estimate for λ_D (`https://cutt.ly/RmBkKKI`.

2. The Debye frequencies of 1-D chain, 2-D square lattice, and 3-D cubic lattices are given by $\omega_D = k_n c_s/a$. $k_1 = \pi$, $k_2 = 2\sqrt{\pi}$, $k_3 = (6\pi^2)^{1/3}$. One obtains an idea about the range of the sound velocities at `https://cutt.ly/hmBkZGX`, which are typically by two orders of magnitude large than the sound velocity in air.

$T_c = T_H$ requires an interaction between condensed matter and magnetic flux tubes carrying dark matter.

1. In the TGD inspired model of living matter [47, 53], the magnetic body (MB) receives information from the biological body (BB) and controls it. For instance, biophotons would be dark photons transformed to ordinary photons.

2. Communication and control would use energy conserving resonant interaction between dark matter associated with the flux tubes and ordinary matter. In particular, sound waves with $h_{eff} = h$ can be transformed to dark photons with $h_{eff} = h_{gr}$ satisfying $E = hf_{high} = \hbar_{eff}f_{low}$, could be example of energy resonance. Living matter is ferroelectric and the transformation of acoustic waves to dark em waves is possible.

3. The resonant transformation of photons to dark photons and back to phonons could give rise to the interaction usually interpreted as a phonon exchange. In the model of cell membrane and EEG, cell membrane sends dark Josephson photons to MB and MB responds by sending dark cyclotron photons absorbed by dark variant of DNA central in TGD inspired model of genetic code [53]. Also acoustic oscillations of cell membrane and DNA are important and also these could participate in the resonance.

2.5.2 TGD based interpretation of the gap energy

The decrease of some kind of binding energy as one approaches T_c from below is highly suggestive. Some kind of binding energy - gap energy ΔE - seems to be involved. At $T = 0$, BCS theory predicts the universal relationship between ΔE and critical temperature T_c

$$\Delta E(T = 0) = 1.1764 T_c \ . \tag{2.16}$$

As one approaches T_c, the gap energy obeys the formula

$$\Delta E(T) = 3.06\sqrt{1 - \frac{T}{T_c}} \ . \tag{2.17}$$

Consider now the TGD inspired interpretation of the gap energy.

1. In the TGD framework ΔE represents the difference $\Delta E = E_B - E_{dark}$ of the binding energy E_B liberated in the formation of the Cooper pair the additional energy of the dark Cooper pair due to $h_{eff} > h$ property (the energy of state as function of h_{eff} increases with h_{eff}). The decrease of E_B with increasing T, which could be caused by weakening of phonon-electron interactions, implies critical temperature.

   For temperatures below T_c, $\Delta E = E_B - E_{dark} < 0$ implies that $h \to h_{eff}$ transition is possible. The surplus energy can be realized as kinetic energy of supra currents. At T_c ΔE vanishes and above T_c $h \to h_{eff}$ transition is impossible.

2. One can however consider a situation in which external energy feed could provide the needed energy. There are situations in which this kind of transitions might be induced thermally or by external energy feed. Indeed, in biology metabolic energy feed would make possible high-Tc superconductivity above Tc.

3. From the gap energy, BCS model predicts the maximal momentum of the Cooper pair as $p_{max}^{1)} = 2m_e\Delta E/p_F$ (in units $c = 1$) allowing to estimate the velocity range for the Cooper pairs of supracurrent.

   In TGD ΔE would go to the longitudinal energy of cyclotron state and transversal cyclotron energy due to the magnetic field of the flux tube. If all energy goes to the momentum, one has $p_{max}^{2)} = \sqrt{4m_e\Delta E}$. This gives for the ratio $p_{max}^{2)}/p_{max}^{1)}$

$$\frac{p_{max}^{2)}}{p_{max}^{1)}} = \sqrt{E_F\Delta} = \sqrt{\frac{E_F}{1.764T_c}} \quad .$$

   For conventional SCs, this ratio is of order $10^3 - 10^4$ since the value of E_F varies in the range 2-10 eV and the value of T_c is in the range $.1 - 1$ meV. This would suggest that cyclotron energy of the Cooper pair with the scale $\hbar_{eff}qB/m_e$ takes most of the energy.

2.5.3 What could be the values of the monopole flux magnetic field and h_{eff}?

In order to say something about the value of B and h_{eff}, some assumptions are needed.

1. The generalization of the Nottale hypothesis [14] to TGD context [29, 28] makes sense. Nottale hypothesis introduces gravitational Planck constant $\hbar_{eff} = \hbar_{gr} = GMm/v_0$, where M in the recent situation is the Earth's mass M_E and v_0 is velocity parameter.

2. The monopole part of the Earth's magnetic field corresponds to the endogenous magnetic field $B_{end} \simeq 2B_E/5 = .2 \times 10^{-4}$ Tesla [19, 20] [47] deduced from the effects of ELF em fields on mammal brain by Blackman and others [15]. The spectrum of B_{end} is assumed to contain also other values, in particular a representation of 12-note scale [52, 53, 54] but this particular value seems to be of special importance.

3. The monopole flux tubes, which carry the field B_{end} are identifiable as gravitational flux tubes mediating gravitational interaction. Whether this is the case or not has remained an open question.

4. Assume that gravitational flux tubes are essential for SC so that quantum gravitation in the TGD sense would be a central element of SC. Therefore SC would not be a mere local condensed matter phenomenon but depend also on M_E and the Earth's gravitational field. Life is also a phenomenon of this kind and the TGD based quantum model for living systems indeed involves high-Tc superconductivity.

Consider now the consequences of these assumptions.

1. For $\hbar_{gr}$, cyclotron energies

$$E_c(\hbar_{gr}) = \hbar_{gr}\frac{qB_{end}}{m} = \frac{GM}{v_0}qB_{end} = \frac{\hbar_{gr}}{\hbar}E_c(\hbar) \qquad (2.18)$$

2. are independent of the mass m of the charged particle. This universality reflects Equivalence Principle.

   Second consequence is that the gravitational Compton length

$$\lambda_{gr} = \frac{\hbar_{gr}}{m} = \frac{GM}{v_0} \qquad (2.19)$$

   is also universal.

3. The model of fountain effect of super-fluidity suggests $v_0 = c/2$ near the surface of Earth. This predicts that λ_{gr} equals to Scwartschild radius r_S which is 9 mm for Earth. All particles would have this gravitational Compton length.

   Also smaller values of v_0 are possible: for instance, $v_0/c \simeq 2^{-11}$ would be true for the inner planets of the Sun [56, 44, 55].

4. $v_0 = c/2$ predicts that in the case of electron

$$\frac{\hbar_{gr}(m)}{\hbar} = \frac{2GMm}{v_0\hbar} = \frac{r_S}{L_c(m)} \quad . \qquad (2.20)$$

   From the value $r_S = 9$ mm for the Earth's Schwartschild radius and the value of electron Compton length $L_c(e) = 2.4 \times 10^{-12}$ m, one obtains

$$\frac{\hbar_{gr}(e)}{\hbar} \simeq .4 \times 10^{10} \quad .$$

   The cyclotron frequency f_c of electron in the endogenous magnetic field B_{end} is $f_c \simeq 6 \times 10^5$ Hz giving for cyclotron energy $E_c(\hbar) = 2.48 \times 10^{-9}$ eV. This gives for

$$E_c(\hbar_{gr}) \simeq 9.3 \text{ eV} \quad ,$$

   which is near to the upper bound of the Fermi energies $E_F(T = 0)$ for electrons in condensed matter.

5. For the Cooper pairs of ions suggested to be crucial for living matter, the same prediction holds true. The same prediction for E_c holds true for bosonic ions. What is interesting is that the prediction would have the same scale as for electrons for neutrinos and their possibly existing Cooper pairs. For neutrinos and neutrons E_c would be replaced by the cyclotron energy in classical Z^0 magnetic fields necessarily accompanying induced Kähler fields at monopole flux tubes.

   The large parity breaking effects in living matter have no convincing explanation in the standard physics framework in living matter. This supports the view about large h_{eff} scaling up also the weak Compton scale. $\hbar_{eff} = \hbar_{gr} = GM_Em/v_0$ with $v_0 = c/2$, the gravitational Compton length would be $\lambda_{gr} = r_S = .9$ cm for all particles including neutrinos and weak bosons. Since dark weak bosons would be effectively massless below this scale, large weak interaction induced parity breaking effects would take place below Λ_{gr}. It is of course not clear whether there exists a mechanism for the formation of neutrino Cooper pairs.

For Sun and inner planets one has $v_0 \simeq 2^{-11}$ and $r_S = 3$ km. This gives $\lambda_{gr} \simeq 2^{10} r_S = 6$ Mm to be compared with the radius $r_S = 6.37$ Mm of Earth. Does this mean that there is a quantum coherent phase in this scale associated with the Earth? For Sun and inner planets one has $v_0 \simeq 2^{-11}$ and $r_S = 3$ km. This gives $\lambda_{gr} \simeq 2^{10} r_S = 6$ Mm to be compared with the radius $r_S = 6.37$ Mm of Earth. Does this mean that there is a quantum coherent phase in this scale associated with the Earth?

2.5.4 TGD based model of Josephson effect

The basic assumption is that the flux tube connection carries a quantum coherent superconducting phase at Josephson junction. In the simplest description, one can apply the Schrödinger equation for the Scrödinger amplitude of Cooper pairs. Therefore the situation reduces to that already considered by Josephson. Is the Hamiltonian just a single particle Hamiltonian for Cooper pairs.

The kinetic part of the Cooper pair Hamiltonian is quadratic in Cooper pair oscillator operators at both sides. The kinetic part becomes linear by coherent state property. Coupling to the vector potential is with charge 2e as in the standard model. Cooper pairs and free electrons move in an external voltage plus helical magnetic field carrying monopole flux giving rise to the "novel" magnetic field.

The covariant constancy condition

$$(\partial_\mu - 2eA_\mu)\psi = 0 \tag{2.21}$$

is satisfied for two coordinates: for the time coordinate (or possibly light-like coordinate) and for the longitudinal coordinate varying along the flow lines of the Beltrami flow. A_μ reduces inb 1-D situation to gradient. Covariant constancy is satisfied at the flux tubes along the helical flux lines and gives Josephson effect in standard manner. The phase is essentially the integral of voltage.

By coherent state property H becomes linear perturbation just like a perturbation of a harmonic oscillator by a periodic force. The effect is non-trivial only in second order perturbation theory. Chemical potential term is not needed at the level of MB.

2.5.5 The 4 anomalies in TGD framework

The article of Koizumi [12] mentions 4 anomalies of the BCS model (no generally accepted model of high-Tc SC exists). Besides the absence of the difference of chemical potentials in the condition defining Josephson frequencies, 3 other anomalies are mentioned. These anomalies do not plague the TGD based model. The basic reason is that Cooper pairs reside at the magnetic flux tubes.

1. There is only one transition temperature in the BCS model of SC whereas high-T_c superconductivity involves 2 transition temperatures. In the TGD framework the first transition temperature leads to a superconductivity but in spatial and time scales (proportional to h_{eff}), which are so short that macroscopic super-conductivity is not possible. In the lower transition temperature h_{eff} increases and the flux tubes reconnect in a stable manner to longer flux tubes. The instability of this phase at critical temperature would be due to the geometric instability of the flux tubes.

2. London moment depends on the real electron mass m_e rather than the effective mass m_e^* of the electron. This effect relates to a rotating magnet. There is a supra current in the boundary region creating the magnetic moment. The explanation is that the electrons resulting from the splitting of Cooper pairs at the flux tubes of magnetic field do not interact with the ordinary condensed matter so that the mass is m_e-

ISSN: 2153-8301   Prespacetime Journal   www.prespacetime.com
Published by QuantumDream, Inc.

3. For SCs of type I, the reversible phase transition from SC to ordinary phase in an external magnetic field does not cause dissipation. One would expect that the splitting of Cooper pairs produces electrons, which continue to flow and dissipate in collisions with the ordinary condensed matter. The reversibility of the phase transition can be understood if the electrons continue to flow at the flux tubes as supracurrents.

4. Magnetic flux tubes also solve the anomaly related to chemical potential: chemical potentials are present but not at the level of magnetic flux tubes so that the erratic calculation gives a correct result in the standard approach.

2.5.6   The basic objection against the TGD based proposal

The basic objection against the TGD based model of superconductivity is that supercurrents flow along monopole flux tubes but an experimental fact is that magnetic field destroys superconductivity. The problem disappears by analyzing the anatomy of magnetic fields in the TGD framework.

1. TGD predicts two kinds of flux tubes carrying Earth's magnetic field B_E with a nominal value of .5 Gauss. This prediciton is quite general. The flux tubes have a closed cross section - this is possible only in TGD Universe, where the space-time is 4-surface in $M^4 \times CP_2$. The flux tubes can have a vanishing Kähler magnetic flux or non-vanishing quantized monopole flux: this has no counterpart in Maxwellian electrodynamics.

   For Earth, the monopole part would correspond to about .2 Gauss - 2/5 of the full strength of B_E.

2. Monopole part needs no currents to maintain it and this makes it possible to understand how the Earth's magnetic field has not disappeared a long time ago. This also explains the existence of magnetic fields in cosmological scales.

   The orientation of the Earth's magnetic field is varying. In the TGD based model, the monopole part plays the role of master. When the non-monopole part becomes too weak, the magnetic body defined by the monopole part changes its orientation. This induced currents refresh the non-monopole part [36]. The standard dynamo model is part of this model.

3. There is an interesting (perhaps more than) analogy with the standard phenomenological description of magnetism in condensed matter. One has $B = H + M$. H field is analogous to the monopole part and the non-monopole part is analogous to the magnetization M induced by H. $B = H + M$ would represent the total field. If this description corresponds to the presence of two kinds of flux tubes, the TGD view about magnetic fields would have been part of electromagnetism from the beginning!

   Flux tubes can also carry electric fields and also for them this kind of decomposition makes sense. Could also the fields D, P, and E have a similar interpretation?

   In the linear model of magnetism, one has $M = \chi H$ and $B = \mu H = (1 + \chi)H$. For diamagnets one has $\chi \leq 0$ and for paramagnets $\chi \geq 0$. Earth would be paramagnetic with $\chi \simeq 3/2$ if the linear model works. χ is a tensor in the general case so that B and H can have different directions.

4. Superconducting phase is a perfect diamagnet so that $B = H + M = 0$. Supra currents generate M, which effectively cancels H. This happens for the interaction of the test particle with the fields H and M, which are at different space-time sheets. In the interaction the test particle touches these space-time sheets and the effects superpose linearly. At the QFT limit this corresponds to the vanishing of B. B does not destroy superconductivity but superconductivity destroys B. In the Meissner effect superconductivity is lost and B is weakened and monopole field H and possible flux tubes of the external field become visible.

3 Summary and conclusions

3.1 Comparison of BPM and TGD inspired model of SC

Consider now the relation of BPM to TGD inspired model of SC.

1. In TGD, the phase factor of complex order parameter would be an exponential of a longitudinal coordinate Φ related to a helical flux along a flux tube serving as a longitudinal coordinate. For closed flux tubes with the shape of a long flattened square, the phase factors at the two flux tubes would be exponentials of the same longitudinal cyclic coordinate. There is no obvious reason for the interpretation as Berry's phase although this interpretation cannot be excluded.

2. In TGD, the "novel magnetism" associated with the Berry phase would correspond to the monopole part of the magnetic field not present in Maxwellian theory. The monopole part plays a central role in TGD inspired quantum biology and also in the model of galaxies and stars [37, 38]. They appear also in the models of hadrons and nuclei and their dark variants leading to a new physics about hadrons and nuclei.

   The flux tubes have closed transversal cross sections and are therefore not possible in Minkowski space. These flux tubes appear in all scales and form a fractal hierarchy.

   Also flux tubes with closed cross section with 2-D homologically trivial projection are possible and carry vanishing magnetic flux as also half flux tubes glued to background 3-surfaces as representation of ordinary flux tubes for which cross section as the topology of disk.

3. The decay of the Beltrami phases could correspond to the decay of a flux tube carrying a Beltrami flow to thinner flux tubes parallel to the original flow. SC would reduce to SC in a shorter scale. The two transitions for high Tc cuprate SCs could correspond to reverse transitions in which flux tubes fused to thicker and longer flux tubes. Low temperature would stabilize longer and thicker flux tubes against splitting to shorter and thinner ones.

It is useful to list the basic differences between BPM and the TGD based model.

1. The authors identify Josephson junction as an insulator. In the TGD framework the junction would consist of superconducting flux tubes accompanied by a parallel structure at the level of ordinary matter, which can be an insulator.

2. In TGD there is a supracurrent of Cooper pairs but it occurs at magnetic flux tubes. Also a supracurrent of electrons is possible.

3. A pair of flux tubes is present in the junction. A reconnection of U-shaped tentacles gives rise to the junctions. Flux tube junctions stabilized have $h_{eff} > h$ and the states have higher energy. The energy liberated in the formation of Cooper pairs provides the energy needed to increase h_{eff}.

4. BPM produces Josephson effect using first order Hamiltonian for thin junctions. For thick junctions a second kind of Josephson effect would result for long junctions.

   In TGD JE does not depend on the length of JE assuming that the junction is accompanied by a magnetic flux tube pair. JE results as a second order effect from the effective Hamiltonian for Cooper pairs which is by coherent state property linear in oscillator operators of Cooper pairs. Situation is essentially the same as in the standard model. Also the mechanism for the formation of Cooper pairs remains the same.

5. BPM predicts chemical potential term. In the TGD framework this is neither predicted nor needed since chemical potential is not needed at the flux tube level. Standard calculation gives a correct result although it is not logically consistent.

3.2 Speculations and questions

The only way to make progress is to speculate and then challenge the speculations by making critical questions. The following represents a list of such speculations and critical questions.

3.2.1 Speculations

Consider first some speculations.

1. The TGD inspired model suggests that SC could be possible also above T_c by using energy feed providing the energy needed to increase the value of h_{eff}. This would be the basic role of metabolism. This could have far reaching technological consequences and also profound implications concerning the creation of artificial life.

   Furthermore, the TGD based model for "cold fusion" [49, 50, 51] led to a reformulation of nuclear physics [38] in which phase transition to dark phase of nuclei has a key role also in the ordinary nuclear reactions as a description of tunnelling phenomenon.

2. In the TGD inspired quantum biology, the cell membrane is identified as a generalized Josephson junction between superconductors assignable to lipid layers of the cell membrane (actually decomposing in a better resolution to membrane proteins acting as Josephson junctions). One can ask what a straightforward application of the basic formulas gives in the case of neuronal membrane.

   One can estimate the gap energy Δ from the formula $\Delta = \hbar\omega_D$ using the already discussed formula $\omega_D = k_n c_s/a$, where k_n depends on the effective dimension of the lattice like system and has values $k_n \in \{3.14, 3.54, 2.66\}$ for $n = 1, 2, 3$. Sound velocity c_s can be replaced with the conduction velocity v of nerve pulses varying in the range $v/c \in [.1, 1] \times 10^6$. The formula would give for $n = 2$ and maximal value $v/c = 10^{-6}$ $E_D = .044$ eV which is in the range of neuronal membrane potentials.

3. The role of $\hbar_{gr}$ and B_{end} in the model would suggest that the SC observed in laboratories is not a mere local condensed matter phenomenon. What happens to SC on Mars? Is the Earth mass replaced with that of Mars and the monopole part B_{end} with its value in Mars? There is evidence that B_{end} is non-vanishing: for instance, Mars has auroras.

4. If the monopole flux tube indeed mediates graviton exchanges, one can wonder whether SC itself is an essentially quantum gravitational phenomenon. Could the attractive interaction between electrons of the Cooper pair be somehow due to gravitation?

   The extremely weak direct gravitational interaction between electrons and nucleons cannot be responsible for the formation of Cooper pairs. One can however argue that Earth takes the role of atomic nuclei in the proposed description. Earth attracts the electrons and causes an effective attraction between them. Could this interaction force the wave functions of the electrons of the Cooper pair with wavelength $\Lambda_{gr} = r_S = 2GM \simeq 9$ mm to overlap and form a quantum coherent state.

   The proposed duality between gauge theories and gravitation, in particular AdS/CFT duality, has a TGD counterpart. The dynamics for the orbits of partonic 2-surfaces and lower-dimensional surface defining a frame for the space-time surface as an analog of soap film [45] would be dual to the dynamics in the interior of the space-time surfaces.

   Could the descriptions in terms of cyclotron photon exchanges and graviton exchanges be dual to each other? Note also that at the fundamental level classical TGD are expressible using only 4 classical field-like variables as a selected subset of imbedding space coordinates. This implies extremely strong constraints between fundamental interactions.

ISSN: 2153-8301   Prespacetime Journal   www.prespacetime.com
Published by QuantumDream, Inc.

3.2.2 Critical questions

Consider now some critical questions.

1. Suppose that Cooper pairs are formed at the level of ordinary matter by interaction with phonons (say) and transferred to MB.

   Q: How can the Cooper pairs survive at MB, where acoustic oscillations mediating interaction with atoms are not present?

   A: The present of the resonant interaction between photons and dark photons would make possible the survival of the Cooper pairs at MB.

   Second option is that the Cooper pairs remain in the ordinary matter and only the electrons are transferred to the flux tubes and the energy liberated in the formation of Cooper pairs makes the transfer energetically possible. Supracurrents would indeed consist of electrons as proposed in [12].

2. I have routinely used the statement "particle resides at magnetic body".

   Q: What does this really mean?

   A: In many-sheeted space-time, the space-time sheets with common 4-D M^4 projection are extremely near to each other and the test particle touches all the sheets. The conclusion in the case of gravitational flux tubes has been that the particle touches all sheets of the many-sheeted magnetic body rather than resides at it.

   Here one must remember that also many-sheetedness with respect to CP_2 is predicted and leads to the proposal that coherent flux tube bundles in M^4 as many-sheeted space-time with respect to CP_2 explain the value of G in terms of CP_2 length squared following from TGD as a prediction [22]. In this case, the test particle does not touch all the sheets unless it has a large value of h_{eff}: $h_{eff} = h_{gr}$ could imply this.

   Q: But doesn't this mean that the particle touches all space-time sheets for all flux tubes?

   A: The generalization of Beltrami hypothesis might actually prevent this. If the M^4 part of the Kähler current is proportional to instanton current and conserved, the M^4 projection of flux tube is 3-D so instanton density vanishes. In this case space-time surfaces have 3-D M^4 projection and are like orbits of membranes and the above argument fails.

   It can of course, also happen that only the sum of M^4 and CP_2 currents has vanishing divergence: in this case the M^4 and CP_2 projections would be 4-D.

   Clearly, the situation is unclear but it is now possible to formulate questions and possible answers precisely.

3. The isotope effect of superconductivity means the proportionality $T_c \propto M^p$, where M is the mass of the isotope. The values of p are near $p = -1/2$. This implies the proportionality $\Delta \propto M^p$.

   Q: Can the $T_c = T_H$ hypothesis be consistent with the isotope effect?

   A: Assume that the scale of dark cyclotron energies determines to a high extent the value of Δ. Cyclotron energies are of the form $E_c = \hbar_{gr} qB/m = (GM/v_0)qB_{end}$. Nottale hypothesis implies that v_0 takes the role of a dimensionless coupling constant strength for gravitation and very probably does not vary [55].

   The local value of B_{end} can however vary and depend on M. This would mean a local variation of the thickness of the flux tube as a response to the contact of the isotope with the isotope. This in turn would cause the local change of the string tension as a sum of densities of the volume and Kähler magnetic energies per unit length.

4 Appendix: General considerations related to the preferred extremals

The following text is based on the updated view about the material from the appendix of [16]. More details can be found in [21, 23, 16].

4.1 Beltrami ansatz and minimal surface ansatz for the preferred extremals of Kähler action

The vanishing of Lorentz 4-force for the induced Kähler field means that the vacuum 4-currents are in a mechanical equilibrium.

1. Lorentz 4-force vanishes for all known solutions of field equations which inspires the hypothesis that all extremals or at least the absolute minima of Kähler action satisfy the condition. The vanishing of the Lorentz 4-force in turn implies local conservation of the ordinary energy momentum tensor. Its vanishing encourages the proposal that Einstein's equations hold true at the Yang-Mills-Einstein limit of TGD.

   The absence of the classical dissipation is highly attractive in the case of supra phases. This condition could be universal and be satisfied below the scales defined by the space-time surface.

   The corresponding condition is implied by Maxwell-Einstein's equations in General Relativity.

2. The hypothesis would mean that the solutions of field equations are what might be called generalized Beltrami fields defining integrable flows serving as candidates for flow lines of the superfluid flow and supracurrents.

3. The hypothesis that Kähler current is proportional to a product of an arbitrary function ψ of CP_2 coordinates and of the instanton current

$$j_I^\alpha = \epsilon^{\alpha\beta\gamma\delta} A_\beta J_{\gamma\delta} \tag{4.1}$$

   solves the 4-D Beltrami condition and reduces to it when electric field vanishes.

   Instanton current has a vanishing divergence for $D_{CP_2} < 4$, and Lorentz 4-force indeed vanishes since the contractions of the gradients of $D < 4$ CP_2 coordinates with a 4-D permutations symbol are involved. Instanton current vanishes for $D = 2$. Note that massless extremals having 2-D CP_2 projection carry non-vanishing light-like current and vanishing instanton current.

   The condition implies that Kähler current can be non-vanishing only if the dimension D_{CP_2} of the CP_2 projection of the space-time surface is less than four so that in the regions with $D_{CP_2} = 4$ (say CP_2 type extremals) Maxwell's vacuum equations are satisfied by the Kähler form.

4. Beltrami fields are known to be extremely complex but highly organized structures and the same is expected to be true for their generalizations (it is not clear whether Kähler current j is always vanishing or light-like).

   An interesting conjecture is that topologically quantized many-sheeted magnetic and Z^0 magnetic Beltrami fields and their 4-D generalizations serve as templates for the helical molecules populating living matter, and explain both chirality selection, the complex linking and knotting of DNA and protein molecules, and even the extremely complex and self-organized dynamics of biological systems at the molecular level.

ISSN: 2153-8301   Prespacetime Journal   www.prespacetime.com
Published by QuantumDream, Inc.

5. Field equations can be reduced to algebraic conditions stating that energy momentum tensor and second fundamental form have no common components (this occurs also for minimal surfaces in string models) and only the conditions stating that Kähler current vanishes, is light-like, or proportional to instanton current, remain and define the remaining field equations. The conditions guaranteeing topologization to instanton current can be solved explicitly. Solutions can be found also in the more general case when Kähler current is not proportional to instanton current (massless extremals). On the basis of these findings there are strong reasons to believe that classical TGD is exactly solvable.

Minimal surface ansatz [45] for the preferred extremals based on 4-D generalization of complex structure to Hamilton-Jacobi structure. It emerges naturally at M^8 level as a prerequisite of $M^8 - H$ duality, and is induce at the level of H by $M^8 - H$ duality [32, 33, 34, 40, 41, 43].

The twistor twistor lift of TGD [27, 30] leads to a concrete action principle at the level of H involving volume term and Kähler action obtained by a dimensional reduction of the Kähler action at the level of twistor space of H. The Kähler action for twistor space exists only in case of H. Therefore there are 3 different views about preferred extremals and they are proposed to be equivalent.

4.2   The dimension of CP_2 projection as classifier for the fundamental phases of matter

The dimension D_{CP_2} of CP_2 projection of the space-time sheet encountered already in p-adic mass calculations classifies the fundamental phases of matter. For $D_{CP_2} = 4$ empty space Maxwell equations hold true. This phase is chaotic and analogous to a demagnetized phase. It might be that only CP_2 type extremals with Euclidean signature of the induced metric and 1-D light-like M^4 projection correspond to this phase.

$D_{CP_2} = 2$ phase is analogous to the ferromagnetic phase: highly ordered and relatively simple. $D_{CP_2} = 3$ is the analog of spin glass and liquid crystal phases, extremely complex but highly organized by the properties of the generalized Beltrami fields. This phase is the boundary between chaos and order and corresponds to life emerging in the interaction of magnetic bodies with bio-matter. It is possible only in a finite temperature interval (note however the p-adic hierarchy of critical temperatures) and characterized by chirality just like life. Both these phases could correspond to SC.

4.3   Connection of Beltrami flows with PCAC hypothesis, massivation, and CP violation

Conserved vector current hypothesis (CVC) and partially conserved axial current hypothesis (PCAC) are essential elements of old-fashioned hadron physics and hold true also in the standard model.

1. The simplest ansatz, which realizes the Beltrami hypothesis, states that the vectorial Kähler current J equals apart from sign $c = \pm 1$ to instanton current I, which is axial current:

$$J = \pm I \ c \ . \tag{4.2}$$

The condition states that only the left or right handed current chiral defined as

$$J_{L/R} = J \pm I \tag{4.3}$$

is non-vanishing. For $c \neq 1$, both J_L and J_R are non-vanishing. Since both right- and left-handed weak currents exist, $c \neq 1$ seems to be a plausible option.

By quantum classical correspondence, these currents serve as space-time correlates for the left- and right-handed fermion currents of the standard model. Note however that induced gamma matrices differ from those of M^4: for instance, they are not covariantly constant but define by field equations a current with a vanishing covariant divergence. Field equations serve as a consistency condition for the modified Dirac action.

2. A more general condition allows c to depend on space-time coordinates. The conservation of J forces conservation of I if the condition $\partial_\alpha c I^\alpha = 0$ is true. This gives a non-trivial condition only in regions with 4-D CP_2 and M^4 projections.

3. The twistor lift of TGD [27, 30] requires that also M^4 has Kähler structure. Therefore J and I and corresponding Kähler gauge potential A have both M^4 part and CP_2 parts and Kähler action K, A_K, J_K, J and I are sums of M^4 and CP_2 parts:

$$
\begin{aligned}
A_K &= A_K(M^4) + A_K(CP_2) \ , \quad &J_K &= J(M^4) + J_K(CP_2) \ , \\
K &= K(M^4) + K(CP_2) \ , \quad &J &= J(M^4) + J(CP_2) \ . \\
I &= I(M^4) + I(CP_2) \ .
\end{aligned}
\tag{4.4}
$$

Only the divergence of I must vanish:

$$
\partial_\alpha I^\alpha = 0 \ .
\tag{4.5}
$$

A possible interpretation is in terms of the 8-D variant of twistorialization by twistor lift [27, 30] requiring masslessness in 8-D sense.

PCAC states that the divergence of the axial current is non-vanishing. This is not in conflict with the conservation of the total instanton current I. PCAC corresponds to the non-conservation $I(CP_2)$, whose non-conservation is compensated by that of $I(M^4)$.

4. For regions with at most 3-D M^4- and CP_2 projections, the M^4- and CP_2 instanton currents have identically vanishing divergence. In these regions the conservation of I is not lost if c has both signs. c could be also position dependent and even differ for $I(M^4)$ and $I(CP_2)$ in these regions.

$D_\alpha I^\alpha$ vanishes for the known extremals. For the simplest CP_2 type extremals and for extremals with 2-D CP_2 projection, I itself vanishes. Therefore parity violation is not possible in these regions. This would suggest that these regions correspond to a massless phase.

5. $D_\alpha I^\alpha \neq 0$ is possible only if both M^4 and CP_2 projections are 4-D. This phase is interpreted as a chaotic phase and by the non-conservation of electroweak axial currents could correspond to a massive phase.

CP_2 type extremals have 4-D projection and for them Kähler current and instanton current vanish identically so that also they correspond to massless phase (M^4 projection is light-like). Could CP_2 type extremals allow deformations with 4-D M^4 projection (DEs)?

The wormhole throat between space-time region with Minkowskian signature of the induced metric and CP_2 type extremal (wormhole contact) with Euclidian signature is light-like and the 4-metric is effectively 3-D. It is not clear whether this allows 4-D M^4 projection in the interior of DE.

The geometric model for massivation based on zitterbewegung of DE provides additional insight [45].

1. $M^8 - H$ duality allows to assign a light-like curve also to DE. For space-time surfaces determined by polynomials (cosmological constant $\Lambda > 0$), this curve consists of pieces which are light-like geodesics.

   Also real analytic functions ($\Lambda = 0$) can be considered and they would allow a continuous light-like curve, whose definition boils down to Virasoro conditions. In both cases, the zigzag motion with light-velocity would give rise to velocity $v < c$ in long length scales having interpretation in terms of massivation.

2. The interaction with $J(M^4)$ would be essential for the generation of momentum due to the M^4 Chern-Simons term assigned with the 3-D light-like partonic orbit. M^4 Chern-Simons term can be interpreted as a boundary term due to the non-vanishing divergence of $I(M^4)$ so that a connection with two views about massivation is obtained. Does the Chern-Simons term come from the Euclidean or Minkowskian region?

I have proposed two models for the generation of matter-antimatter asymmetry. In both models, CP breaking by M^4 Kähler form is essential. Classical electric field induces CP breaking. CP takes self-dual (E, B) to anti-self-dual $(-E, B)$ and self-duality of $J(M^4)$ does not allow CP as a symmetry.

1. In the first model the electric part of $J(M^4)$ would induce a small CP breaking inside cosmic strings thickened to flux tubes inducing in turn small matter-antimatter asymmetry outside cosmic strings. After annihilation this would leave only matter outside the cosmic strings.

2. In the simplest variant of TGD only quarks are fundamental particles and leptons are their local composites in CP_2 scale [39, 42].

   Both quarks and antiquarks are possible but antiquarks would combine leptons as almost local 3-quark composites and presumably realized CP_2 type extremals with the 3 antiquarks associated with the partonic orbit. I should vanish identically for the DEs representing quarks and leptons but not for antiquarks and antileptons.

   Could the number of DEs with vanishing I be smaller for antiquarks than for quarks by CP breaking and could this induce leptonization of antiquarks and favor baryons instead of antileptons? Could matter-antimatter asymmetry be induced by the interior of DE alone or by its interaction with the Minkowskian space-time region outside DE.

In the standard model also charged weak currents are allowed. Does TGD allow their space-time counterparts? CP_2 allows quaternionic structure in the sense that the conformally invariant Weyl tensor has besides $W_3 = J(CP_2)$ also charged components $W_\pm$, which are however not covariantly constant [57]. One can assign to $W_\pm$ analogs of Kähler currents as covariant divergences and also the analogs of instanton currents. These currents could realize a classical space-time analog of current algebra.

Received August 4, 2021; Accepted December 4, 2021

References

[1] Lakthakia A. *Beltrami Fields in Chiral Media*, volume 2. World Scientific, Singapore, 1994.

[2] Fradkin E. *Field Theories of Condensed Matter Physics*. Cambridge University Press, 2013.

[3] Ghrist R Etnyre J. An index for closed orbits in Beltrami field, 2001. Available at: `http://arxiv.org/abs/math/010109`.

[4] Marsh GE. *Helicity and Electromagnetic Field Topology*. World Scientific, 1995.

[5] Bogoyavlenskij OI. Exact unsteady solutions to the Navier-Stokes equations and viscous MHD equations. *Phys Lett A*, pages 281–286, 2003.

[6] Scientists Detect 'Fingerprint' of High-temperature Superconductivity Above Transition Temperature. *Science Daily*, 2009. Available at: http://www.sciencedaily.com/releases/2009/08/090827141338.htm.

[7] Tranquada JM Emery VJ, Kivelson SA. Stripe phases in high-temperature superconductors. *Perspective*, 96(16), 1999. Available at: http://www.pnas.org/cgi/reprint/96/16/8814.pdf.

[8] Deutcher G. Andreev–saint-james reflections: A probe of cuprate superconductors guy deutscher. *Rev. Mod. Phys.*, 77(109), 2005. Available at: https://arxiv.org/abs/cond-mat/0409225.

[9] Zaanen J. Why high Tc is exciting?, 2005. Available at: http://www.lorentz.leidenuniv.nl/research/jan_hitc.pdf.

[10] Zaanen J. Superconductivity: Quantum Stripe Search. *Nature*, 2006. Available at: http://www.lorentz.leidenuniv.nl/~jan/nature03/qustripes06.pdf.

[11] Zaanen J. Watching Rush Hour in the World of Electrons. *Science*, 2007. Available at: http://www.ilorentz.org/~jan/perspstripes.pdf.

[12] Hiroyasu Koizumi. Superconductivity by Berry connection from many-body wave functions: revisit to Andreev-Saint-James reflection and Josephson effect. *Journal of Superconductivity and Novel Magnetism*, 2021. Available at: https://arxiv.org/pdf/2103.00805.pdf.

[13] Buchanan M. Mind the pseudogap. *Nature*, 409:8–11, 2001.Available at: http://www.physique.usherbrooke.ca/taillefer/Projets/Nature-409008.pdf.

[14] Nottale L Da Rocha D. Gravitational Structure Formation in Scale Relativity, 2003. Available at: http://arxiv.org/abs/astro-ph/0310036.

[15] Blackman CF. *Effect of Electrical and Magnetic Fields on the Nervous System*, pages 331–355. Plenum, New York, 1994.

[16] Pitkänen M. About Strange Effects Related to Rotating Magnetic Systems . In *TGD and Fringe Physics*. Available at: http://tgdtheory.fi/pdfpool/Faraday.pdf, 2006.

[17] Pitkänen M. Bio-Systems as Super-Conductors: part I. In *Quantum Hardware of Living Matter*. Available at: http://tgdtheory.fi/pdfpool/superc1.pdf, 2006.

[18] Pitkänen M. Bio-Systems as Super-Conductors: part II. In *Quantum Hardware of Living Matter*. Available at: http://tgdtheory.fi/pdfpool/superc2.pdf, 2006.

[19] Pitkänen M. Quantum Model for Bio-Superconductivity: I. In *TGD and EEG*. Available at: http://tgdtheory.fi/pdfpool/biosupercondI.pdf, 2006.

[20] Pitkänen M. Quantum Model for Bio-Superconductivity: II. In *TGD and EEG*. Available at: http://tgdtheory.fi/pdfpool/biosupercondII.pdf, 2006.

[21] Pitkänen M. About Preferred Extremals of Kähler Action. In *Physics in Many-Sheeted Space-Time: Part I*. Available at: http://tgdtheory.fi/pdfpool/prext.pdf, 2019.

[22] Pitkänen M. About the Nottale's formula for h_{gr} and the possibility that Planck length l_P and CP_2 length R are related. In *Hyper-finite Factors and Dark Matter Hierarchy: Part I*. Available at: http://tgdtheory.fi/pdfpool/vzerovariableG.pdf, 2019.

ISSN: 2153-8301   Prespacetime Journal   www.prespacetime.com
Published by QuantumDream, Inc.

[23] Pitkänen M. Basic Extremals of Kähler Action. In *Physics in Many-Sheeted Space-Time: Part I*. Available at: http://tgdtheory.fi/pdfpool/class.pdf, 2019.

[24] Pitkänen M. Recent View about Kähler Geometry and Spin Structure of WCW . In *Quantum Physics as Infinite-Dimensional Geometry*. Available at: http://tgdtheory.fi/pdfpool/wcwnew.pdf, 2014.

[25] Pitkänen M. WCW Spinor Structure. In *Quantum Physics as Infinite-Dimensional Geometry*. Available at: http://tgdtheory.fi/pdfpool/cspin.pdf, 2006.

[26] Pitkänen M. Could ZEO provide a new approach to the quantization of fermions? Available at: http://tgdtheory.fi/public_html/articles/secondquant.pdf., 2020.

[27] Pitkänen M. Some questions related to the twistor lift of TGD. In *Towards M-Matrix: Part II*. Available at: http://tgdtheory.fi/pdfpool/twistquestions.pdf, 2019.

[28] Pitkänen M. Quantum Astrophysics. In *Physics in Many-Sheeted Space-Time: Part II*. Available at: http://tgdtheory.fi/pdfpool/qastro.pdf, 2019.

[29] Pitkänen M. TGD and Astrophysics. In *Physics in Many-Sheeted Space-Time: Part II*. Available at: http://tgdtheory.fi/pdfpool/astro.pdf, 2019.

[30] Pitkänen M. The Recent View about Twistorialization in TGD Framework. In *Towards M-Matrix: Part II*. Available at: http://tgdtheory.fi/pdfpool/smatrix.pdf, 2019.

[31] Pitkänen M. Quantitative model of high T_c super-conductivity and bio-super-conductivity. Available at: http://tgdtheory.fi/public_html/articles/newsupercond.pdf., 2015.

[32] Pitkänen M. Does $M^8 - H$ duality reduce classical TGD to octonionic algebraic geometry?: part I. Available at: http://tgdtheory.fi/public_html/articles/ratpoints1.pdf., 2017.

[33] Pitkänen M. Does $M^8 - H$ duality reduce classical TGD to octonionic algebraic geometry?: part II. Available at: http://tgdtheory.fi/public_html/articles/ratpoints2.pdf., 2017.

[34] Pitkänen M. Does $M^8 - H$ duality reduce classical TGD to octonionic algebraic geometry?: part III. Available at: http://tgdtheory.fi/public_html/articles/ratpoints3.pdf., 2017.

[35] Pitkänen M. Two new findings related to high Tc super-conductivity. Available at: http://tgdtheory.fi/public_html/articles/supercondnew.pdf., 2018.

[36] Pitkänen M. Maintenance problem for Earth's magnetic field. Available at: http://tgdtheory.fi/public_html/articles/Bmaintenance.pdf., 2015.

[37] Pitkänen M. Cosmic string model for the formation of galaxies and stars. Available at: http://tgdtheory.fi/public_html/articles/galaxystars.pdf., 2019.

[38] Pitkänen M. Solar Metallicity Problem from TGD Perspective. Available at: http://tgdtheory.fi/public_html/articles/darkcore.pdf., 2019.

[39] Pitkänen M. SUSY in TGD Universe. Available at: http://tgdtheory.fi/public_html/articles/susyTGD.pdf., 2019.

[40] Pitkänen M. A critical re-examination of $M^8 - H$ duality hypothesis: part I. Available at: http://tgdtheory.fi/public_html/articles/M8H1.pdf., 2020.

[41] Pitkänen M. A critical re-examination of $M^8 - H$ duality hypothesis: part II. Available at: http://tgdtheory.fi/public_html/articles/M8H2.pdf., 2020.

[42] Pitkänen M. Can one regard leptons as effectively local 3-quark composites? `https://tgdtheory.fi/public_html/articles/leptoDelta.pdf.`, 2021.

[43] Pitkänen M. Is $M^8 - H$ duality consistent with Fourier analysis at the level of $M^4 \times CP_2$? `https://tgdtheory.fi/public_html/articles/M8Hperiodic.pdf.`, 2021.

[44] Pitkänen M. Three alternative generalizations of Nottale's hypothesis in TGD framework. `https://tgdtheory.fi/public_html/articles/MDMdistance.pdf.`, 2021.

[45] Pitkänen M. What could 2-D minimal surfaces teach about TGD? `https://tgdtheory.fi/public_html/articles/minimal.pdf.`, 2021.

[46] Pitkänen M and Rastmanesh R. Homeostasis as self-organized quantum criticality. Available at: `http://tgdtheory.fi/public_html/articles/SP.pdf.`, 2020.

[47] Pitkänen M and Rastmanesh R. The based view about dark matter at the level of molecular biology. Available at: `http://tgdtheory.fi/public_html/articles/darkchemi.pdf.`, 2020.

[48] Pitkänen M and Rastmanesh R. Aging from TGD point of view. `https://tgdtheory.fi/public_html/articles/aging.pdf.`, 2021.

[49] Pitkänen M. Cold Fusion Again . Available at: `http://tgdtheory.fi/public_html/articles/cfagain.pdf.`, 2015.

[50] Pitkänen M. Cold fusion, low energy nuclear reactions, or dark nuclear synthesis? Available at: `http://tgdtheory.fi/public_html/articles/krivit.pdf.`, 2017.

[51] Pitkänen M. Could TGD provide new solutions to the energy problem? Available at: `http://tgdtheory.fi/public_html/articles/proposal.pdf.`, 2020.

[52] Pitkänen M. Geometric theory of harmony. Available at: `http://tgdtheory.fi/public_html/articles/harmonytheory.pdf.`, 2014.

[53] Pitkänen M. How to compose beautiful music of light in bio-harmony? `https://tgdtheory.fi/public_html/articles/bioharmony2020.pdf.`, 2020.

[54] Pitkänen M. Is genetic code part of fundamental physics in TGD framework? Available at: `https://tgdtheory.fi/public_html/articles/TIH.pdf.`, 2021.

[55] Pitkänen M. Questions about coupling constant evolution. `https://tgdtheory.fi/public_html/articles/ccheff.pdf.`, 2021.

[56] Pitkänen M. Time reversal and the anomalies of rotating magnetic systems. Available at: `https://tgdtheory.fi/public_html/articles/freereverse.pdf.`, 2021.

[57] Pitkänen M. The Geometry of CP_2 and its Relationship to Standard Model. 2010. Available at: `https://tgdtheory.fi/pdfpool/appendb.pdf.`

Essay

Possible Explanation of Long Cosmic Spinning Filaments

Matti Pitkänen [1]

Abstract

A long filament with length of order 10^8 ly characterizing the sizes of large cosmic voids has been studied. The filament consists of galaxies and the surprising finding is that besides moving along the filament, the galaxies associated with the filaments spin around the filament axis. This finding suggests a network of filaments of length of order 10^8 ly and thickness of order 10^6 ly intersecting at the nodes formed by large galaxy clusters. The larger the masses at the ends of the filament are, the larger the spin is. How angular momentum is generated is the problem. In Newtonian and General Relativistic frameworks it is very difficult to imagine any plausible mechanism. TGD suggests a mechanism in which the compensating angular momentum associated with visible objects is associated with the dark matter associated with the cosmic strings and the flux tubes resulting as they thicken. Dark matter would therefore have a fundamental role in the gravitational dynamics in all scales.

1 Introduction

The inspiration for writing this article came from a highly interesting popular article (`https://cutt.ly/inMODTT`) providing new information about the cosmic filaments (thanks to Jebin Larosh for the link). The popular article tells about the article published in Nature [3] (`https://cutt.ly/HnMOGcP`) and telling about the work of a team led by Noam Libeskind.

What has been studied is a long filament with length of order 10^8 ly characterizing the sizes of large cosmic voids. The filament consists of galaxies and the surprising finding is that besides moving along the filament, the galaxies associated with the filaments spin around the filament axis.

This finding suggests a network of filaments of length of order 10^8 ly and thickness of order 10^6 ly intersecting at nodes formed by large galaxy clusters. The larger the masses at the ends of the filament are, the larger the spin is.

How angular momentum is generated is the problem. The problem is quite general and is shared by both Newtonian and General Relativistic Universes. The natural assumption is that angular momentum vanishes in the original situation. Angular momentum conservation requires a generation of compensating angular momentum. This should happen in the case of all rotating structures. Already the case of galaxies is problematic but if the length scale of the structure is 10^8 ly, the situation becomes really difficult.

Gravitationally bound states have as a rule angular momentum preventing gravitational collapse but how the angular momentum is generated in a process believed to be a concentration of a homogeneous matter density to astrophysical objects? The basic problem is that the Newtonian description relies on scalar potential so that the field lines of the Newtonian gravitational field are never closed. It is difficult to imagine mechanisms for the generation of angular momentum by rotation. In the GRT based description gravi-magnetic fields, which are rotational, emerge but they are extremely weak. The proposal is that tidal forces could generate angular momentum but the generation of angular momentum remains poorly understood.

---

[1]Correspondence: Matti Pitkänen `http://tgdtheory.fi/`. Address: Rinnekatu 2-4 8A, 03620, Karkkila, Finland. Email: matpitka6@gmail.com.

ISSN: 2153-8301   Prespacetime Journal   www.prespacetime.com
Published by QuantumDream, Inc.

2   TGD view about the angular momentum generation

Could one understand the recent finding, and more generally, the generation of angular momentum, in the TGD framework? What raises hope is that in the TGD framework Kähler magnetic fields, whose flux tubes can be regarded as space-time quanta, are key players of dynamics in all scales besides gravitation.

2.1   Cosmic strings as carriers of dark matter and energy

The basic difference between GRT and TGD are cosmic strings and flux tubes resulting from their thickening. Cosmic strings are preferred extremals which are space-time surfaces with 2-D string world sheet as M^4 projection and complex surface of CP_2 as CP_2 projection.

1. The presence of the long filaments is one of the many pieces of support for the fractal web of cosmic strings thickened to flux tubes predicted by TGD. The scale is the scale of large voids 10^8 ly forming a kind of honeycomb like structure. The density of matter would be fractal in the TGD Universe [12, 13] (`http://tgdtheory.fi/public_html/articles/meco.pdf` and `http://tgdtheory.fi/public_html/articles/galaxystars.pdf`).

2. Long cosmic string has a gravitational potential proportional to $1/\rho$, ρ the transverse distance. This predicts a flat velocity spectrum for the stars rotating around the galaxy. No dark matter halo is needed. The model contains only a single parameter, string tension, and also this can be understood in terms of the energy density of the cosmic string. The motion along the string is essentially free motion which allows to distinguish the model from the halo model. In fact, the article [3] reports linear motion along the filament.

   Amusingly, the same day that I learned about the spinning filaments, I learned about a new evidence for the absence of the galactic halo from a popular article (`https://cutt.ly/MnMOI7F` telling about the article by Shen et al [2] (`https://cutt.ly/HnMOPNA`).

2.2   Compensating angular moment as angular momentum of dark matter at cosmic string

Consider now the problem of how the compensating angular momentum is generated as visible matter starts to rotate.

In the TGD framework the picture is just the opposite.

1. The basic assumption of the Newtonian and GRT based models for the generation of angular momentum is that all astrophysical objects are formed by a condensation of matter along perturbations of the mass density. The flow of mass occurs from long scales to short scales.

2. Cosmic strings are the basic objects present already in primordial cosmology [9, 6, 5]. Long cosmic strings form tangles along them in a local thickening, which gives rise to flux tubes [12, 13, 14]. This involves the decay of dark energy and matter at cosmic string to ordinary matter around them as the string tension is reduced in a phase transition decreasing the coefficient of the volume term present in the action besides Kähler action as predicted by twistor lift of TGD [7, 10]. This parameter corresponds to length scale dependent cosmological constant Λ.

   Λ depends on p-adic length scale $L_p \propto \sqrt{p}$, $p \simeq 2^k$ and satisfies $\Lambda(k) \propto 1/L^2(k)^2$. $\Lambda(k)$ approaches zero in long p-adic length scales characterizing the transversal size of flux tubes. This solves the cosmological constant problem. The thickness $d \sim L(k)$ of the flux tube, which is rather small, determines the string tension. To $L(k)$ there is associated a long p-adic length scale which is of order size of observed cosmology if $d \sim L(k)$ is of order of 10^{-4} meters, which happens to be the size of a large neuron.

3. The phase transitions reducing Λ reduce string tension are analogous to the decay of the inflaton field vacuum energy to ordinary matter. Now inflaton field vacuum energy is replaced with the dark energy and matter associated with the thickening cosmic string. Each phase transition is accompanied by an accelerated expansion. The period known as inflation in stanaard cosmology is the first phase transition of this kind. The recent accelerated expansion would correspond to a particular period of this kind and will eventually slow down.

What could happen in the decay of the energy of a flux tube tangle of a cosmic string to visible matter?

1. The visible matter resulting in the decay of the cosmic string must start to rotate around the cosmic string since otherwise it would fall back to the cosmic string like matter into a blackhole. The cosmic string must somehow generate a spin compensating the angular momentum of the visible matter.

2. One should understand angular momentum conservation. Generation of visible matter with angular momentum is possible only if the dark cosmic string is helical or becomes (increasingly) helical in the phase transitions. The angular momentum would be accompanied by the longitudinal motion along the string: this motion has been observed for the filaments [3] .

   The helical structure could be present from the beginning or be generated during the decay of energy of the cosmic string leading to the local thickenings to flux tube giving rise to galaxies as tangles along a long cosmic string. Also the dark matter and energy at the cosmic string already have angular momentum so that the dark matter that transforms to visible matter would inherit this angular momentum.

   The reported correlation between the masses at the ends of the filament and the spin of the filament [3] could be understood if the masses at the ends are formed from the dark energy and mass of the filament having angular momentum. The amount of spin and mass at the ends would be the larger, the longer the decay process has lasted.

3. The identification of the galaxies as tangles along long cosmic strings explains the flatness of the galactic velocity spectrum. Galaxy rotates and also now the angular momentum conservation is the problem. The simplest solution is that the cosmic string portions between the tangles generate the angular momentum opposite to that of the visible matter.

   This would happen not only for the portions of cosmic string between galaxies but also those between stars in the galactic tangle. Stars would be flux tube spaghettis and the angular momentum of the star would be compensated by the angular momentum associated with the helical cosmic string continuing outside the star and connecting it to other stars.

The illustration of the popular article brings in mind a DNA double strand and inspires a consideration of an alternative, perhaps unnecessarily complex, model.

1. Suppose one has a double helix of cosmic strings, call them Alice and Bob. Two stellar objects can form a gravitationally stable state only if relative rotation is present. This would be true also for a cosmic double strand to prevent gravitational collapse in 2-D sense.

2. Alice could remain a cosmic string and thus dark so that we would not see it. Bob would thicken to a flux tube and produce ordinary matter as galaxies as ordinary matter realized tangles along it. The matter would inherit the angular momentum the dark matter and energy producing it already has. The string tension of Bob would be reduced in this process. Of course, both Alice and Bob could have tangles along them. The experiments however support the view that spin direction is the same along the filament.

3. If the helical pair of cosmic strings is actually a closed loop in which the second strand is a piece of the same string, the motion of matter along strands is automatically in opposite directions and spins are opposite. The rotational motion as a stabilizer of a gravitationally bound state is transformed to a helical motion. The problem is however why only the other strand decays to ordinary matter (in the case of ordinary DNA there is an analogous problem due to the passivity of the second strand).

2.3   Is quantum gravitation cosmic scales involved?

There is an interesting connection to atomic physics suggesting that quantum effects are associated with gravitationally bound dark matter even in astrophysical scales.

1. The basic problem was that the electron should radiate its energy and fall into the atomic nucleus. The Bohr model of the atom solved the problem and non-radiating stationary states prevented the infrared catastrophe. Also in the gravitational case something similar is expected to happen for gravitational interaction.

2. The Bohr model of solar system [8, 4], originally introduced by Nottale [1], relies on the notion of gravitational Planck constant $hbar_{gr} = GMm/\beta_0$ predicts angular momentum quantization [11, 15].

3. Angular momentum quantization as multiples of $\hbar_{gr}$ could occur also for the matter rotating around the cosmic string. In the case of the filament, the mass M could be replaced with the mass of the cosmic string (or possibly several of parallel cosmic strings) and m could correspond to the mass of a galaxy rotating around it. The velocity parameter $\beta_0 = v_0/c$ has a spectrum of values [15] proposed to come as inverse integers.

Received June 28, 2021; Accepted December 4, 2021

References

[1] Nottale L Da Rocha D. Gravitational Structure Formation in Scale Relativity, 2003. Available at: http://arxiv.org/abs/astro-ph/0310036.

[2] Shen Z et al. A Tip of the Red Giant Branch Distance of 22.1 1.2 Mpc to the Dark Matter Deficient Galaxy NGC 1052 DF2 from 40 Orbits of Hubble Space Telescope Imaging. *The Astrophysical Journal Letters*, 914(1), 2021. Available at: https://iopscience.iop.org/article/10.3847/2041-8213/ac0335.

[3] Wang P et al. Possible observational evidence for cosmic filament spin. *Nature Astronomy*, 2021. Available at: https://www.nature.com/articles/s41550-021-01380-6.

[4] Pitkänen M. About the Nottale's formula for h_{gr} and the possibility that Planck length l_P and CP_2 length R are related. In *Hyper-finite Factors and Dark Matter Hierarchy: Part I*. Available at: http://tgdtheory.fi/pdfpool/vzerovariableG.pdf, 2019.

[5] Pitkänen M. Cosmic Strings. In *Physics in Many-Sheeted Space-Time: Part II*. Available at: http://tgdtheory.fi/pdfpool/cstrings.pdf, 2019.

[6] Pitkänen M. More about TGD Inspired Cosmology. In *Physics in Many-Sheeted Space-Time: Part II*. Available at: http://tgdtheory.fi/pdfpool/cosmomore.pdf, 2019.

[7] Pitkänen M. Some questions related to the twistor lift of TGD. In *Towards M-Matrix: Part II*. Available at: http://tgdtheory.fi/pdfpool/twistquestions.pdf, 2019.

[8] Pitkänen M. TGD and Astrophysics. In *Physics in Many-Sheeted Space-Time: Part II*. Available at: `http://tgdtheory.fi/pdfpool/astro.pdf`, 2019.

[9] Pitkänen M. TGD and Cosmology. In *Physics in Many-Sheeted Space-Time: Part II*. Available at: `http://tgdtheory.fi/pdfpool/cosmo.pdf`, 2019.

[10] Pitkänen M. The Recent View about Twistorialization in TGD Framework. In *Towards M-Matrix: Part II*. Available at: `http://tgdtheory.fi/pdfpool/smatrix.pdf`, 2019.

[11] Pitkänen M. TGD view about coupling constant evolution. Available at: `http://tgdtheory.fi/public_html/articles/ccevolution.pdf.`, 2018.

[12] Pitkänen M. TGD view about quasars. Available at: `http://tgdtheory.fi/public_html/articles/meco.pdf.`, 2018.

[13] Pitkänen M. Cosmic string model for the formation of galaxies and stars. Available at: `http://tgdtheory.fi/public_html/articles/galaxystars.pdf.`, 2019.

[14] Pitkänen M. Solar Metallicity Problem from TGD Perspective. Available at: `http://tgdtheory.fi/public_html/articles/darkcore.pdf.`, 2019.

[15] Pitkänen M. Questions about coupling constant evolution. `https://tgdtheory.fi/public_html/articles/ccheff.pdf.`, 2021.

Essay

Does the Notion of Polynomial of Infinite Order Make Sense?

Matti Pitkänen [1]

Abstract

$M^8 - H$ duality relates number theoretical and geometric visions of physics in the TGD framework. At the level of M^8 polynomials with rational coefficients would define the space-time surfaces as the "roots" of their complexified octonionic continuations. The basic dynamical principle states that they have associative normal spaces. In principle, also an analytic function with rational Taylor coefficients is possible and can give rise to transcendental extensions. A longstanding question has been whether it makes sense to talk about polynomials of infinite degree. It turns out that if the polynomials of infinite degree exist, they must correspond to composites for an infinite number of polynomials. This follows from the fact that both finite and infinite Galois groups must be profinite so that an infinite Galois group is a Galois group of ...extensions of extensions.....of rationals. The example in which the polynomials of form $P = P \circ R$ where Q is an infinite composite of a single polynomial Q vanishing at origin and having it as a critical point has as a basin of attraction a set having Julia set as boundary. All points in the basin of attraction for origin are roots at the limit.All points in the basin of attraction for origin are roots at the limit so that an algebraic completion of rationals to complex numbers would result.

1   Introduction

TGD provides motivations for the question whether the notion polynomial of infinite degree could make sense. In the following I consider this question from the point of view of physicist and start from the vision about physics as generalized number theory.

1.1   Background and motivations for the idea

$M^8 - H$-duality ($H = M^4 \times CP_2$) states that space-time surfaces defined as 4-D roots of complexified octonionic polynomials so that they have quaternionic normal space, can be mapped to 4-surfaces in H [4, 5, 9].

The octonionic polynomials are obtained by algebraic continuation of ordinary real polynomials with rational coefficients although one can also consider algebraic coefficients.

This construction makes sense also for analytic functions with rational (or algebraic) coefficients. For the twistor lift of TGD, cosmological constant Λ emerges via the coefficient of a volume term of the action containing also Kähler action. This leads to an action consisting of Kähler action with both CP_2 and M^4 terms having very interesting and physically attractive properties, such as spin glass degeneracy. $\Lambda = 0$ would correspond to an infinite volume limit making the QFT description possible as an approximate description. Also the thermodynamic limit could correspond to this limit.

Irreducible polynomials of rational coefficients give rise to algebraic extensions characterized by the Galois group and these notions are central in adelic vision.

I do not know of any deep reason preventing analytic functions with rational Taylor coefficients. These would make possible transcendental extensions. For instance, the product $\prod_p (e^x - p))$ for some subset of primes p would give as roots transcendental numbers $log(p)$. The Galois group would be however trivial although the extension is infinite. Second example is provided by trigonometric functions $sin(x)$ and $cos(x)$ with roots coming as multiples of $n\pi$ and $(2n + 1)\pi/2$. This might be necessary in order to have

---

[1]Correspondence: Matti Pitkänen http://tgdtheory.fi/. Address: Rinnekatu 2-4 8A, 03620, Karkkila, Finland. Email: matpitka6@gmail.com.

Fourier analysis. The translations by a multiple of π for x act permuted roots but do not leave rational numbers rational so that the interpretation as a Galois group is not possible so that also now Galois group would be trivial.

A long standing question has been whether there exist analytic functions which could be regarded as polynomials of infinite order by posing some conditions to the Taylor coefficients. If so, one might hope that the notion of Galois group could make sense also now, and one might perhaps obtain a unified view about transcendental extensions of rationals.

1. For polynomials as roots of octonionic polynomials space-time surfaces are finite and located inside finite-sized causal diamond (CD).

   In the TGD Universe cosmological constant Λ depends on the p-adic length scale and approaches zero at infinite length scale. At the $\Lambda = 0$ limit, which corresponds also to QFT and thermodynamical limits, space-time surfaces would have infinite size. Only Kähler action with M^4 and CP_2 parts and having ground state degeneracy analogous to spin glass degeneracy would be present.

2. The octonionic algebraic continuations of analytic functions with rational coefficients and subject to restrictions guaranteeing that the notion of prime function makes sense, would define space-time surfaces as their roots.

3. Prime analytic functions defining space-time surfaces would in some sense be polynomials of infinite degree and could be even characterized by the Galois group. For real polynomials complex conjugations for the roots is certainly this kind of symmetry.

   These functions should have Taylor series at origin, which is a special point for octonionic polynomials with rational (or perhaps even algebraic) coefficients. The selection of origin as a preferred point relates directly to the condition eliminating possible problems due to the loss of associativity and commutativity.

   The prime property is possible only if the set of these polynomials fails to have a field property (so that the inverse of any element would be well-defined) since for fields one does not have the notion of prime. The field property is lost if the allowed functions vanish at origin so that one cannot have a Taylor series at origin and the inverse diverges at origin.

   The vanishing at origin guarantees that the functional composite $f \circ g$ of f and g has the roots of g. Roots are inherited as algebraical complexity as a kind of evolution increases. In TGD inspired biology, the roots of polynomials are analogous to genes and the conservation of roots in the function composition would be analogous to the conservation of genes.

1.2   Attractor basin of fractal as set of roots

It turns out that if the polynomials of infinite degree exist, they must correspond to composites for an infinite number of polynomials. This follows from the fact that both finite and infinite Galois groups must be profinite so that an infinite Galois group is a Galois group of ...extensions of extensions.....of rationals.

The example in which the polynomials of form $P = P \circ R$ where Q is an infinite composite of single polynomial Q vanishing at origin and having it as a critical point has as a basin of attraction a set having Julia set as boundary. All points in the basin of attraction for origin are roots at the limit.

Profiniteness suggests an interpretation of this set in terms of p-adic topology or a product of a subset of p-adic number fields somehow determined by the number theoretic properties of Q. Algebraic completions of p-adic topology could also be in question. p-Adic number fields are indeed profinite and as additive groups can act as infinite Galois groups permuting the zeros. The action of p-adic translations could leave the basin of attraction invariant.

ISSN: 2153-8301   Prespacetime Journal   www.prespacetime.com
Published by QuantumDream, Inc.

2 What it is to be a polynomial of infinite degree?

In the following the conditions on the notion of polynomial of infinite degree are discussed.

2.1 Conditions for the prime analytic function

Could one make anything concrete about this idea? What kind of functions f could serve as analogs of polynomials of infinite degree with transcendental roots. The question whether any analytic function with rational coefficients vanishing at origin can have a possibly unique decomposition to prime analytic functions will not be discussed in the sequel?

1. Suppose that the analytic prime decomposes to a product over monomials $x - x_i$ with transcendental roots x_i such that the Taylor series has rational coefficients. This requires an infinite Taylor series.

2. One obtains an infinite number of conditions. Each power x^n in f has a rational coefficient f_n equal to the sum over all possible products $\prod_{k=1}^{n} x_{i_k}$ of n transcendental roots x_{i_k}. This gives an infinite number of conditions and each condition involves an infinite number of roots. If the number N of transcendental roots is finite as it is for polynomials, each term involves a finite number of products and the conditions imply that the roots are algebraic. The number of transcendental roots must therefore be infinite. At least formally, these conditions make sense.

3. The sums of products are generalized symmetric functions of transcendental roots and should have rational values equal to x_n. This generalizes the corresponding condition for ordinary polynomials. Symmetric functions for S_n have S_n as a group of symmetries. For a Galois extension of a polynomial of order n, the Galois group is a subgroup of S_n. This suggests that the Galois group is a subgroup of S_∞. S_∞ as the simple A_∞ as a subgroup of even permutations. The simple groups are analogs of primes for finite groups and one can hope that this is true for infinite and discrete groups [8].

There are infinitely many ways to represent an algebraic extension in terms of a polynomial and the same is true for transcendental extensions with the rationality condition.

1. Consider a general decomposition of the polynomial of an infinite order to a product of monomials with roots spanning the transcendental extension. Could a suitable representation of extension as an infinite polynomial allow rational coefficients f_n for the function $\sum f_n x^n$ defined by the infinite product?

2. f_n is the sum over all possible products of roots obtained by dropping n different roots from the product of all roots which should be finite and equal to one for the generalization of monic polynomials. Therefore there is an infinite sum of terms, which are inverses of finite products and therefore transcendental but one can hope that the infinite number of the summands allows the rationality condition to be satisfied.

2.2 Profinite groups and Galois extensions as inverse limits

Infinite groups indeed appear as Galois groups of infinite extensions. Absolute Galois groups, say Galois groups of algebraic numbers, provide the basic example.

1. There exists a natural topology, known as Krull topology, which turns Galois group to a profinite group (totally disconnected, Hausdorff topological group) (`https://en.wikipedia.org/wiki/Profinite_group`), which is also Stone space (`https://en.wikipedia.org/wiki/Stone_space`).

2. Profinite groups are not countably infinite but are effectively finite just as hyper-finite factors of type II_1 are finite-dimensional: they appear naturally in the TGD framework [3, 2]. Profinite groups

have Haar measure giving them a finite volume. Profinite groups behave in many respects like finite groups (compact groups also behave in this manner as far representations are considered). Profiniteness is possessed by products, closed subgroups, and the coset groups associated with the closed normal subgroups.

3. Every profinite infinite group is a Galois group for an infinite extension for some field K but one cannot control which field K is realized for a given profinite group [1]. Additive p-adics groups and their products appear as Galois groups of an infinite extension for some field K. The Galois theory of infinite field extensions involves profinite groups obtained as Galois groups for the inverse limits of finite field extensions $..F_n \to F_{n+1} \to$.

4. This kind of iterated extensions are of special interest in the TGD framework and an infinite extension would be obtained at the limit [7]. The naive expectation is that the polynomial of infinite degree is a limit of a composite $...P_n \circ P_{n-1}.. \circ P_1$ of rational polynomials. The number of infinite extensions obtained in this manner would be infinite.

   An interesting question is under what conditions the limiting infinite polynomial exists as an analytic function and whether the Taylor coefficients are rational or in some extension of rationals. The naive intuition is that the inverse limit preserves rationality.

5. The identification as the iterate $...P_n \circ P_{n-1} \circ P_1$ is indeed suggestive. Infinite cyclic extension defined at the limit by the polynomial x^N, $N = \infty$, to be discussed below, has this kind of interpretation. The Galois group of this kind of extension is however not simple.

   Remark: The polynomials in question are not irreducible: the composite of N polynomials has x^N as a factor.

6. Is the infinite-D extension obtained as an inverse limit transcendental or algebraic? In the TGD framework the condition that the polynomial $P_1 \circ P_2$ has the roots of P_1 as roots implies the loss of the field property of analytic functions making the notion of analytic prime possible. The roots of the infinite polynomial contain all roots of finite polynomials appearing in the sequence. This would suggest that the extension is not transcendental. Giving up the property $P_i(0) = 0$ also leads to a loss of root inheritance.

For finite-dimensional Galois extensions, there exists an infinite number of polynomials generating the extension and one can consider families of extensions parametrized by a set of rational parameters. The Galois group does not change under small variations of parameters [8]. If the inverse limit based on an infinite composite of polynomials makes sense, the situation could be the same for possibly existing rational polynomials of infinite order? The study of infinite Galois groups could provide insights on the problem.

2.3   Could infinite extensions of rationals with a simple Galois group exist?

Simple Galois groups have no normal subgroups and are of special interest as the building bricks of extensions by functional composition of polynomials. The infinite Galois groups obtained as inverse limit have however an infinite hierarchy of normal subgroups and simple argument suggests that the extensions are algebraic. Could infinite-D transcendental extensions defined by an analytic function with rational coefficients and with a simple infinite Galois group, exist?

   If inverse limit is essential for profiniteness for infinite groups, then simple infinite groups are excluded as Galois groups. Indeed, the topology of an infinite simple group G cannot be profinite. The Krull topology has as a basis for open sets all cosets of normal subgroups H of finite index (the number of cosets gH is finite). Simple group has no normal subgroups except a trivial group consisting of a unit element and the group itself. The only open sets would be the empty set and G itself.

ISSN: 2153-8301   Prespacetime Journal   www.prespacetime.com
Published by QuantumDream, Inc.

In fact, there is also a theorem stating that every Galois group is profinite (see `https://cutt.ly/wQ2W1Of`). All finite groups are profinite in discrete topology. This theorem however excludes infinite simple Galois groups. If one allows only polynomials with P(0)=0, the conservation of algebraic roots suggests that infinite polynomials with transcendental roots are not possible.

The condition for the failure of the field property however leaves the iterates of polynomials for which only the highest polynomial in the infinite sequence of functional compositions vanishes at origin. These infinite polynomials could have transcendental roots.

2.4 Two examples

In the following two examples are consider to test whether the notion of a polynomial of infinite order might work.

2.4.1 Infinite cyclic extensions

The natural question is whether the transcendental roots be regarded as limits of roots for a polynomial with rational coefficients at the limit when the degree N approaches infinity. The above arguments suggest that the limits involve an infinite function composition.

Consider as an example cyclic extension defined by a polynomial X^N, which can be regarded as a composite of polynomials x^{p_i} for $\prod p_i = N$. This is perhaps the simplest possible extension than one can imagine.

1. The roots are now powers of roots of unity. The notion of the root of unity as $e^{i2\pi/N}$ does not make sense at the limit $N \to \infty$. One can however consider the roots $e^{i2\pi M/N}$ and its powers such that the limit $M/N \to \alpha$ is irrational. The powers of $exp(in\alpha)$ give a dense subset of the circle S^1 consisting of irrational points. Note that one obtains an infinite number of extensions labelled by irrational values of α.

2. The polynomial should correspond to the limit $P_N(x) = x^N - 1$, $N \to \infty$. For each finite value of N, one has $P_N(x) = \prod_{n=1}^{N}(x - U^n) - 1$, $U = e^{i2\pi/N}$. The reduction to $P = x^N - 1$ follows from the vanishing of all terms involving lower powers of x than x^N.

3. If these conditions hold true at the limit $N \to \infty$, one obtains the same result. The coefficient of x^N equals to 1 trivially. The coefficient of x^{N-1} is the sum over all roots and should vanish. This is also assumed in Fourier analysis $\sum_n e^{i\alpha n} = 0$ for irrational α. For $\alpha = 0$ the sum equals to $N = \infty$ identified as Dirac delta function. The lower terms give conditions expected to reduce to this condition. This can be explicitly checked for f_1

4. The Galois group is in this case the cyclic group $U_{\infty,\alpha}$ defined by the powers of U_α.

2.4.2 Infinite iteration yields contimuum or roots

The iterations of polynomials define an $N \to \infty$ limit, which can be handled mathematically whereas for an arbitrary sequence of polynomials in the functional composition it is difficult to say anything about the possible emergence of transcendental roots. Note however that the $lim_{N\to\infty}(1 + 1/N)^N = e$ shows that transcendentals can appear as limits of rationals. I have considered iterations of polynomials and approach to chaos from the point of view of $M^8 - H$ duality in [6].

Consider polynomials $P_N = Q_N \circ R$, where R with $Q(0) = 0$ is fixed polynomial and $Q_N = Q^{\circ N}$ is the N:th integrate of some irreducible polynomial Q with $Q(0) = 0$ and $dQ/dz(0) = 0$. Origin is a fixed critical point of Q and the attractor towards which the points in the attractor basin of origin end up in the iteration and become roots of P_∞ and are roots at this limit. For the real points in the intersection of the positive real axis and attractor basin are roots at this limit so that one has a continuum of roots. The set of roots consists of a continuous segment $[0, T)$ and a discrete set coming from the Julia set defining the boundary of the attractor basin.

Profiniteness suggests an interpretation of this set in terms of p-adic topology or a product of a subset p-adic topologies somehow determined by the number theoretic properties of Q. p-Adic number fields are indeed profinite and as additive groups can act as infinite Galois groups permuting the zeros. The action of p-adic translations could indeed leave the basin of attraction invariant.

In TGD framework these roots correspond to values of M^4 time (or energy!) in M^8 mapped to the actual time values in H by $M^8 - H$ duality. I have referrred to them as "very special moments in the life of self" with a motivation coming from TGD inspired theory of consciousness [4, 5]. One might perhaps say that at this limit subjective time consisting of these moments becomes continuous in the interval $[0, T]$.

Received September 6, 2021; Accepted December 5, 2021

References

[1] Waterhouse WC. Profinite groups are Galois groups. *Proceedings of the American Mathematical Society*, 42(2):639–640, 1974. Available at: `https://www.jstor.org/stable/2039560`.

[2] Pitkänen M. Evolution of Ideas about Hyper-finite Factors in TGD. In *Hyper-finite Factors and Dark Matter Hierarchy: Part II*. Available at: `http://tgdtheory.fi/pdfpool/vNeumannnew.pdf`, 2019.

[3] Pitkänen M. Was von Neumann Right After All? In *Hyper-finite Factors and Dark Matter Hierarchy: Part I*. Available at: `http://tgdtheory.fi/pdfpool/vNeumann.pdf`, 2019.

[4] Pitkänen M. A critical re-examination of $M^8 - H$ duality hypothesis: part I. Available at: `http://tgdtheory.fi/public_html/articles/M8H1.pdf.`, 2020.

[5] Pitkänen M. A critical re-examination of $M^8 - H$ duality hypothesis: part II. Available at: `http://tgdtheory.fi/public_html/articles/M8H2.pdf.`, 2020.

[6] Pitkänen M. Could quantum randomness have something to do with classical chaos? Available at: `http://tgdtheory.fi/public_html/articles/chaostgd.pdf.`, 2020.

[7] Pitkänen M. The dynamics of SSFRs as quantum measurement cascades in the group algebra of Galois group. Available at: `http://tgdtheory.fi/public_html/articles/SSFRGalois.pdf.`, 2020.

[8] Pitkänen M. About the role of Galois groups in TGD framework. `https://tgdtheory.fi/public_html/articles/GaloisTGD.pdf.`, 2021.

[9] Pitkänen M. Is $M^8 - H$ duality consistent with Fourier analysis at the level of $M^4 \times CP_2$? `https://tgdtheory.fi/public_html/articles/M8Hperiodic.pdf.`, 2021.

Exploration

Some Questions about Coupling Constant Evolution

Matti Pitkänen [1]

Abstract

In this article questions related to the hierarchy of Planck constants and p-adic coupling constant evolution (CCE) in the TGD framework are considered:

1. Is p-adic length scale hypothesis (PLS) correct in this recent form and can one deduce this hypothesis or its generalization from the basic physics of TGD defined by Kähler function of the "world of classical worlds" (WCW)? The fact, that the scaling of the roots of polynomial does not affect the algebraic properties of the extension strongly suggests that p-adic prime does not depend on purely algebraic properties of EQ. In particular, the proposed identification of p as a ramified prime of EQ could be wrong. Number theoretical universality suggests the formula $exp(\Delta K) = p^n$, where ΔK is the contribution to Kähler function of WCW for a given space-time surface inside causal diamond (CD).

2. The understanding of p-adic length scale evolution is also a problem. The "dark" CCE would be $\alpha_K = g_K^2/2h_{eff} = g_K^2/2nh_0$, and the PLS evolution $g_K^2(k) = g_K^2(max)/k$ should define independent evolutions since scalings commute with number theory. The total evolution $\alpha_K = \alpha_K(max)/nk$ would induce also the evolution of other coupling strengths if the coupling strenghts are related to α_K by Möbius transformation as suggested.

3. The formula $h_{eff} = nh_0$ involves the minimal value h_0. How could one determine it? p-Adic mass calculations for $h_{eff} = h$ lead to the conclusion that the CP_2 scale R is roughly $10^{7.5}$ times longer than Planck length l_P. Classical argument however suggests $R \simeq l_P$. If one assumes $h_{eff} = h_0$ in the p-adic mass calculations, this is indeed the case for $h/h_0 = (R(CP_2)/l_P)^2$. This ratio follows from number theoretic arguments as $h/h_0 = n_0 = (7!)^2$. This gives $\alpha_K = n_0/kn$, and perturbation theory can converge even for $n = 1$ for sufficiently long p-adic length scales. Gauge coupling strengths are predicted to be practically zero at gravitational flux tubes so that only gravitational interaction is effectively present. This conforms with the view about dark matter.

4. Nottale hypothesis predicts gravitational Planck constant $\hbar_{gr} = GMm/\beta_0$ ($\beta_0 = v_0/c$ is velocity parameter), which has gigantic values. Gravitational fine structure constant is given by $\alpha_{gr} = \beta_0/4\pi$. Kepler's law $\beta^2 = GM/r = r_S/2r$ suggests length scale evolution $\beta^2 = xr_S/2L_N = \beta_{0,max}^2/N^2$, where x is proportionality constant, which can be fixed. Phase transitions changing β_0 are possible at $L_N/a_{gr} = N^2$ and these scales correspond to radii for the gravitational analogs of the Bohr orbits of hydrogen. p-Adic length scale hierarchy is replaced by that for the radii of Bohr orbits. The simplest option is that β_0 obeys a CCE induced by α_K. This picture conforms with the existing applications and makes it possible to understand the value of β_0 for the solar system, and is consistent with the application to the superfluid fountain effect.

1 Introduction

In this article questions related to the notions of the p-adic CCE and hierarchy of Planck constants will be considered.

---

[1]Correspondence: Matti Pitkänen http://tgdtheory.fi/. Address: Rinnekatu 2-4 8A, 03620, Karkkila, Finland. Email: matpitka6@gmail.com.

1.1 How p-adic primes are determined?

p-Adic length scale (PLS) hypothesis plays a central role in TGD in all length scales. For instance, it makes it possible to use simple scaling arguments to deduce quantitative predictions for the masses of new particles predicted by TGD.

PLS hypothesis states that the size scales of space-time surfaces correspond to PLSs $L_p = \sqrt{p}R(CP_2)$. The additional hypothesis is $p \simeq m^k$, $m = 2, 3, ...$ a small prime. The success of p-adic mass calculations [18] supports $p \simeq 2^k$ hypothesis [22] seriously. There also exists empirical evidence for a possible generalization to small primes, in particular $m = 3$, in biology [14, 15].

The physical and mathematical identification of the origin of the p-adic prime p defining the PLS is however a problem.

The proposal has been that the p-adic prime p defining the PLS corresponds to a ramified prime of the extension of rationals (EQ) associated with the polynomial defining space-time region in M^8 picture. Ramified primes appear as factors of the discriminant of the polynomial defining EQ. I have not been able to find any really convincing explanation for why p should correspond to a ramified prime so that p-dic prime might emerge in some other manner.

In p-adic thermodynamics Boltzmann weights $exp(-E/T)$ must be replaced with $p^{L_0/T}$, where L_0 is scaling generator. The exponent $\Omega = exp(K)$ of the Kähler function K of WCW defines vacuum functional. Could Ω be number-theoretically universal and thus exist as a p-adic number for some prime p determining naturally the PLS. This is the case if one has $\Omega = p^n$, n integer.

As such, this idea does not make sense but one consider a subsystem defined by sub-CD defining self in zero energy ontology (ZEO) based theory of consciousness [36, 46] [26]?

p-Adic prime defines naturally the scale of CD for trivial extension of rationals and this scale is scaled up by factor n for an extension of dimension n. This also conforms with the assumption that p-adic CCE and "dark" CCE are independent.

1.2 Trying to understand p-adic CCE

TGD leads to a number theoretic vision about CCE [32]. Number theoretic universality plays a key role in this picture. CCE certainly involves the hierarchy of extensions of rationals (EQs) possibly involving non-rational extensions by roots of e, which induce finite extensions of p-adics. It would be nice if the EQ alone would determine the values of the coupling constants.

1. The starting point is that the continuous CCE with respect to length scale reduces to a discrete PLS evolution with respect to L_p, $p \simeq 2^k$. There is also dark evolution with respect to $n = h_{eff}/h_0$. These evolutions are separate since the scaling of the roots of the polynomial do not affect the purely algebraic properties of the extension. The natural assumption is that these evolutions factorize so that one has $\alpha_K = g_K^2(p)/2h_{eff}$.

2. p-Adic CCE would be roughly logarithmic with respect to L_p. The observation that α is near $\alpha = 1/137$ for p-adic length scale $L(137)$ suggests that for α_K defining the fundamental coupling strength one has

$$\alpha_K = \frac{g_K^2(max)}{2kh_{eff}} \quad .$$

Since $1/\alpha_K(137) = 137$ is prime for ordinary matter with $h_{eff} = h$, one must have

$$\frac{g_K^2(max)}{2h} = 1 \quad .$$

giving $h = g_K^2(max)/2$. The value h need not however be the minimal value h_0 of h_{eff} since one can have $h = n_0 h_0$ $\alpha_K(max) = 2n_0$ so that one can write

$$\alpha_K = \frac{1}{knn_0} \ . \tag{1.1}$$

$n_0 > 1$ would mean that the ordinary matter would be actually dark in the sense that the order of the extension of rationals associated with the ground state would be n_0.

For h_0 that value of α_K could be so large that the perturbation series does not converge except in very long length scales for which k is expected to be large. Exotic phases with $h_{eff} < h$ could become possible in these scales.

1.3 How p-adic prime is defined at the level of WCW geometry?

The p-adic prime p should emerge from the dynamics defined by Kähler function.

1. The Kähler function K of the "world of classical worlds" (WCW), or more generally the generalization of $exp(K)$ to a vacuum functional possibly involving also a genuine state dependent part is a central quantity concerning scattering amplitudes. Suppose that one can consider a subsystem defined by CD and the contribution ΔK from CD to K.

   Number theoretical universality suggests that the exponential $exp(\Delta K)$ or its appropriate generalization exists in all p-adic number fields or at least in an extension of the p-adic number field corresponding to the p-adic prime p. Could this condition fix p dynamically?

2. Suppose that for some prime p one can write

$$e^{\Delta K} = p^{\frac{\Delta K}{log(p)}}$$

   such that $\Delta K/log(p)$ is integer. The exponential would be a power of p just as the p-adic analog of Boltzmann weight in p-adic thermodynamics [18]. This would select a unique p-adic prime p defining the PLS and this prime need not be a ramified prime. In p-adic thermodynamics [18] $X = \Delta K/log(p)$ has interpretation as an eigenvalue of the scaling generator L_0 of conformal algebra and one can even consider the possibility that there is a connection.

1.4 What about the evolution of the gravitational fine structure constant?

Nottale hypothesis [11] predicts gravitational Planck constant $\hbar_{gr} = GMm/\beta_0$ ($\beta_0 = v_0/c$ is velocity parameter), which has gigantic values so that the above picture fails. Gravitational fine structure constant is given by $\alpha_{gr} = \beta_0/4\pi$.

Kepler's law $\beta^2 = GM/r = r_S/2r$ suggests length scale evolution $\beta^2 = xr_S/2L_N = \beta_{0,max}^2/N^2$, where x is proportionality constant, which can be fixed. Phase transitions changing β_0 are possible at $L_N/a_{gr} = N^2$ and these scales correspond to radii for the gravitational analogs of the Bohr orbits of gravitational Bohr atom. PLS hierarchy is replaced by that for the radii of Bohr orbits.

What could be the interpretation of N? The safest assumption is that the CCE of β_0 is analogous to that of the other coupling constants and induced from that of α_K.

1.5 What is the minimal value of h_{eff}?

The formula $h_{eff} = nh_0$ involves the minimal value h_0 of h_{eff}. The simplest explanation for the findings of Randell Mills [12] is that one has $h = 6h_0$. h_0 could be also smaller [27].

What is the value of h_0? A possible answer to this question came from the observation made already during the first 10 years of TGD. The observation was that the imbeddings of spherically symmetric stationary metrics (see the Appendix) suggest that CP_2 radius R is of order Planck length l_P rather than by factor about $10^{7.5}$ longer. Could one have $h = n_0 h_0$, $n_0 \sim 10^{7.5}$ so that the ordinary matter would be actually dark?

CP_2 radius would be Planck length apart from numerical constant not far from unity. The p-adic mass calculations would give correct results for $h_{eff} = h_0$. R could be interpreted as $R^2 = n_0 l_P^2$. The perturbative expansion for $h_{eff} < h$ would not converge except in long p-adic length scales, where the p-adic evolution reduces the value of α_K.

Gauge coupling strengths are predicted to be practically zero at gravitational flux tubes with very large h_{eff} so that only gravitational interaction is effectively present. This conforms with the view about dark matter.

2 Number theoretical universality of vacuum functional and p-adic CCE

The Kähler geometry of WCW is defined by a Kähler function $K(X^4(X^3))$ identified as the action of preferred extremal consisting of volume term and Kähler action. The vacuum functional is of form $\Omega = exp(K + iS)$. Here K is the real Kähler function and S is the counterpart of real action in the path integral of QFTs.

$exp(iS)$ could be interpreted as a dynamical part of vacuum functional, which depends on state rather than being "God-given". The reason why this would be the case would be that it is possible. For $exp(K)$ there is no choice since the Kähler geometry of WCW is expected to be unique merely from its existence as already in the case of loop spaces [8].

Number theoretical universality is a challenge for this general picture.

1. In the p-adic context the notion of WCW geometry is highly questionable. The integration associated with definition of volume term and Kähler action is the tough problem.

   This has inspired the proposal that the exponent of the action completely disappears from the scattering amplitudes. This indeed happens in quantum field theory based on path integral around stationary point.

2. The classical nondeterminism suggests a weaker formulation. The sum over the contributions of stationary points would be replaced by integral over preferred extremals consisting of 3-surfaces at PB plus sum over the paths of the tree resulting from classical non-determinism.

   The sum over the paths of the tree-like structure remains in the superposition of amplitudes for sub-CD and it might be possible to define the deviation $\Delta K + i\Delta S$ of the action for each of them and separate $exp(\Delta K + i\Delta S)$ from the entire exponent of action, which would therefore disappear from the expression of the scattering amplitudes for given X^3 and given CD. Otherwise, the knowledge of the entire WCW Kähler function would be needed.

   A possible interpretation is in terms of a decomposition to an unentangled tensor product corresponding to sub-CD and its environment so that one can separate the physics inside sub-CD from that of environment and code it by $exp(\Delta K + i\Delta S)$.

3. The simplest option, very probably too simple, would be that one has $\Delta K = 0$. Kähler function would be same for all paths of the tree and one would obtain a discretized analog of path integral. This would require that all the branches of the tree have same value of action. This does not however require the same value of volume and Kähler action separately.

   It will be foud that $\Delta K \neq 0$ assuming that $exp(\Delta K)$ reduces to an integer power of p for some prime identifiable as p-adic prime defining the PLS, is more interesting option since it would reduce p-adic thermodynamics to the level of WCW and also allow to the understand of PLS evolution of couping constants.

The number theoretical existence of the phases $exp(i\Delta S)$ would require that they belong to the EQ defining the space-time region inside CD.

4. This picture suggests that the number theoretically universal part is associated with the sub-CDs and with the discrete physics of the tree-like structure whereas the Kähler function for 3-surfaces would be defined only in real framework. This would neatly separate the physics of sensory and Boolean cognition as something number theoretically universal from the physics proper, so to say.

   Since conscious experience gives all information about physics, one can ask whether the adelic physics associated with various sub-selves could together be enough to represent all that is representable from the physics proper. This could result as somekind of limiting case (EQ approaches algebraic numbers).

If this view is correct, then one expects that various notions shared by QFTs and TGD, in particular CCE, could have number theoretic descriptions as indeed suggested [32]. In the sequel I will discuss some speculations in this framework.

2.1   The recent view about zero energy ontology

Zero energy ontology (ZEO) [26] [36, 46] plays a key role in the formulation of TGD based quantum measurement theory.

1. The concept of causal diamond (CD) is central. CD serves as a correlate for the perceptive field of conscious entity: this in the case that one has sub-CD so that the space-time surfaces inside CD continue outside it.

   The scale size scale of the CD identifiable as the temporal distance T between its tips could be proportional the p-adic prime p at the lowest level of dark matter hierarchy and to np at dark sectors. p-Adic length scales L_p characterizing the sizes scale of 3-surfaces are proportional to $\sqrt{p}$ and the proposal is that the relation between T and L_p is same as the relationship between diffusion time T and the root mean square distance R travelled by diffusion.

2. The twistor lift of TGD predicts that the action principle defining space-time surfaces is the sum of a volume term characterized by length scale dependent cosmological constant Λ and Kähler action and induced from 6-D Kähler action whose existence fixes the imbedding space uniquely to $M^4 \times CP_2$. The reason is that the required Kähler structure exists only for the twistor spaces of M^4, E^4, and CP_2 [9].

3. The recent progress in the understanding of zero energy ontology (ZEO) [46] leads to rather detailed view about the dynamics of the space-time surfaces inside sub-CD.

   Space-time surfaces are analogs of soap films spanned by a frame having the 3-surfaces at its ends located at the boundary of CD as fixed part of frame and the dynamically generated parts of frame in the interior of CD. Outside the frame preferred extremal is an analog of a complex surface and a simultaneous extremal of both volume term and Kähler action since the field equations reduce to conditions expressing the analogy of holomorphy [23, 25]. The field equations reduce to contractions of tensors of type (1,1) with tensors of type (2,0)+(0,2) and are therefore trivially true.

   The minimal surface property fails at the frame, and only the full field equations are true. The divergences of isometry currents associated with volume term have delta function singularities which however cancel each other to guarantee field equations and conservation laws. This is expected to give rise to a failure of determinism, which is however finite in the sense that the space-time surfaces associated with given 3-surface X^3 at the passive boundary of CD (PB) form a finite set which is a tree-like structure (for a full determinism only single space-time surface as analog of Bohr orbit would be realized). Therefore the non-determinism of classical dynamics for a fixed X^3 is extremely

ISSN: 2153-8301   Prespacetime Journal   www.prespacetime.com
Published by QuantumDream, Inc.

simple and quantum dynamics and classical dynamics are very closely related since quantum states are superpositions of the paths of the tree.

4. One also ends up to quite precise identification sub-CD or space-time surface inside sub-CD as a correlate of perceptive field of a conscious entity. The essential element of the picture that for sub-CD the 3-surface X^3 at PB is fixed but due to the non-determinism the end at active boundary (AB) is not completely fixed and there is finite non-determinism in the state space defined by superpositions of the paths of the tree.

   For the highest level in the hierarchy of CDs associated with self, the space-time surfaces inside CD do not continue outside it and this CD God-like entity, whose dynamics is not restricted by the boundary conditions.

This view provides additional perspectives on discreteness of adelic physics unifying the physics of sensory experience and cognition [30, 31].

1. Discreteness is essential in the number theoretic universality since in these case real structures and their p-adic counterparts correspond naturally to each other. This has led to the notion of cognitive representation as a set of points of space-time surface with preferred imbedding space coordinates having values in the EQ defined by the polynomial defining the space-time surface in complexified M^8 and mapped to H by $M^8 - H$ duality [38, 39]. The finite-dimensionality of the state space associated with the tree structure conforms with this vision.

2. Discreteness is natural for the dynamics of concious experience and cognition. Mental images as sub-selves correspond to the sub-CDs inside CD. Sub-CDs are naturally located at the loci of non-determinism defined by the fixed part of the frame dynamically and generated frames in the interior and at AB.

   Attention would fix the 3-surfaces at the PB of a sub-CD as a perceptive sub-field and all CDs in the hierarchy would be fixed in this manner. The loci of non-determinism would serve as targets of attention. Sensory perception, memory recall, and other functions would reduce to directed attention inside CD.

   Fermionic degrees of freedom at boundaries of CD are are additional discrete degrees of freedom and responsible for Boolean cognition whereas the discrete dynamics of frame would correspond to sensory experience and sensory aspects of cognition.

3. This picture inspires the question whether the number theoretically universal parts of adelic physics might relate to the physics due to the non-determinism in the interior of sub-CD. This physics would be basically the physics that can be observed. This would mean enormous simplification.

   This idea is not new. The amazing success of p-adic thermodynamics based mass calculations [18] could be understood if p-adic physics is seen as a physics of cognitive representation of real number based physics.

In the sequel some speculations are discussed by taking the above picture as a basis.

2.2 Number theoretical constraints on $exp(\Delta K)$

Number theoretical universality suggests that the exponents $exp(\Delta K + i\Delta S)$ for X^4 inside sub-CD is well-defined at least for some p-adic number fields or their extensions.

It has been already found that number theoretical universality requires that the phases $exp(i\Delta S)$ belong to the EQ associated with the space-time surfaces considered.

The condition that the phase is a root of unity is more general than the condition of semiclassical approximation of wave mechanics stating that the action is quantized as a multiple of Planck constant

h. The analog of this condition would imply $exp(i\Delta S) = 1$. This quantization condition would make S obsolete.

What about the number theoretical universality of $exp(\Delta K)$? One can consider three options.

1. p-Adic exponent function $exp(x)$ exists if the p-adic norm of x is smaller than 1. The problem is that the p-adic exponent function and its real counterpart behave very differently [21]. In particular, $exp(x)$ is not periodic. Integer powers of e^p are however ordinary p-adic number by its Taylor series and roots of e define finite-D extensions of p-adic number fields. Therefore $exp(\Delta K)$ could make sense as an integer power for a root of e.

   If ΔK is integer, $exp(\Delta K)$ exists p-adically for primes p dividing ΔK.

2. Also $p^{\Delta K/log(p)}$ could exist p-adically if $\Delta K/log(p)$ is integer. This implies strong conditions. ΔK must be of form $\Delta K = log(p)m$, m integer. If ΔK corresponds to Kähler function of WCW, p is fixed and would define the sought-for preferred p-adic prime p defining the PLS.

3. Since the powers p^n converge to zero for $n \to \infty$, one can formally replace $exp(\Delta K)$ with $exp(\Delta K) = p^{\Delta K)/log(p)}$ and require that the exponent is an integer. The replacement of the ordinary Boltzman weights with powers of p is indeed carried out in p-adic thermodynamics [18]. This suggests that the Boltzman factors of p-adic thermodynamics reduce to exponents $p^{\Delta K}$ at the level of WCW.

3 Hierarchy of Planck constants, Nottale's hypothesis, and TGD

3.1 Nottale's hypothesis

Nottale's hypothesis [11] and its generalization to TGD [17, 19] has non-relativistic and relativistic forms.

1. The non-relativistic formula for $\hbar_{gr}$ as given by the Nottale's formula

$$\hbar_{gr} = \frac{GMm}{\beta_0} \ , $$
$$\alpha_{gr} = \frac{GMm}{4\pi\hbar_{gr}} = \frac{\beta_0}{4\pi} \ . \tag{3.1}$$

   The formula makes sense only $\hbar_{gr}/\hbar > 1$.

2. The relativistically invariant formula for $\hbar_{gr}$ reads for four momenta $P = (M, 0)$ $p = (E, p_3)$ as:

$$\hbar_{gr} = \frac{GP \cdot p}{\beta_0} = \frac{GME}{\beta_0} = \frac{r_s E}{2\beta_0} \ , \tag{3.2}$$

   where r_s is Schwartschild radius. Adelic physics implies that momentum components belong to an extension of rationals defining the adele so that the spectrum of E and of $\hbar_{gr}$ are discretized.

3.1.1 Nottale's hypothesis and biology

Nottale's hypothesis involves a lot of uncertainties also at the conceptual level. Hence it is important to see whether basic facts from TGD inspired biology support the Nottale's hypothesis.

1. The cyclotron frequencies in an "endogenous" magnetic field $B_{end} = 2B_E/5$, where $B_E = .5$ Gauss is the nominal value of the Earth's magnetic emerge in the explanation of the findings of Blackman and other [16] showing that ELF photons have effects on vertebrate brain. B_{end} is assigned with the monopole flux tubes of B_E. Also lower and higher values of B_{end} can be considered and the models of hearing [?] and genetic code [?] suggests that the values of B_{end} correspond to the notes of 12-note scale. This suggests that also the Z^0 magnetic field might be involved.

2. Biophoton energies are in visible and UV range and in the TGD based model they are assumed to result in the transformations of dark photons with much smaller frequency but same energy to ordinary photons. For instance, photons with 10 Hz frequency can transform to biophotons. By $E = h_{eff}f$, requires $h_{eff} = h_{gr}$. The implication is that cyclotron energies do not depend on particle mass. Furthermore, Schwartschild radius $r_S = .9$ cm of Earth defines universal gravitational Compton length for $\beta_0 = 1/2$.

Assume that h_{gr} corresponds to Earth mass and $\beta_0 = 1/2$ and consider cyclotron states in $B_{end} = .2$ Gauss.

1. The value of $r = h_{gr}/\hbar$ for proton is given as the ratio r_s/L_p, where L_p is the Compton radius of proton. This gives $r = .833 \times 10^{13}$. For ions with mass number A the value of r is scaled to Ar.

2. What is the cyclotron energy associated with the 10 Hz frequency in this case? The energy of a photon with frequency f is for $h_{gr}(m_p)$ given by $E_c/eV = r \times 1.24 \times (f/(3 \times 10^{14}Hz)$. Proton's cyclotron frequency is $f_c = 300$ Hz in B_{end} and corresponds to 10 eV, which is in the UV region and rather large.

3. All cyclotron frequencies of charged particles correspond to $E_c = 10$ eV cyclotron energy, which seems rather large. If h_{gr} is reduced by factor $1/4$ as required to explain the findings of Mills at least partially, the cyclotron energy becomes 2.5 eV, which is in the visible range. Scaling by factor $1/2$ gives cyclotron energy 5 eV in UV.

4. Smaller values of E_c would require smaller fields. The Z_0 charge of proton is roughly a fraction $1/50$ of its em charge and since Kähler field contributes also to Z^0 field one would obtain energy about .2 eV in the IR region.

10 Hz alpha frequency which is of special interest concerning understanding of conscious experience and it is interesting to look for concrete numbers.

1. $f = 10$ Hz is alpha frequency and the cyclotron frequency $f_c = 10$ eV Fe^{2+} ion with mass number $A = 56$. Fe^{2+} ions play a central role in biology.

2. For $f = 10$ Hz the energy $h_{gr}(m_p)$ (proton) is .333 eV to be compared with the metabolic energy currency $\sim .5$ eV and is below the visible range.

3. In the TGD inspired biology, 3 proton units represent dark genetic codons and for $h_{gr}(3m_p)$ the energy corresponds to $E \times 1$ eV, which is still slightly below the visible range [48, 41, 43]. In the dark variant of double DNA strand parallel to the ordinary double strand, the 2 dark codons form a pair by the dark variant of the base pairing so that one has effective $A06$ and $E = 2$ eV, which corresponds to red light.

4. The energy $E = 2$ eV of the codon pair for $f = 10$ Hz corresponds formally to $A = 6$ and would characterize ^6Li. Litium's cyclotron frequency is around $f_c = 50$ Hz is known to have biological significance. Li is used in the treatment of depression [?]. One might imagine that the coupling of Lithium to dark codon pairs might be involved.

5. For higher mass numbers, the energies for 10 Hz and $\hbar_{gr}(Am_p)$ belong to the UV region. For oxygen one with $A = 16$ has $E = 5.3$ eV, which could correspond to some important molecular transition energy. Molecular bond dissociation energies (`https://cutt.ly/3QoZxY9`) vary in the range .03 -10 eV. O-H, O=O ad O=CO bond energies are somewhat above 5 eV. The idea indeed is that the transformation of dark photons to ordinary bio-photons allows a control of molecular biochemistry.

6. DNA codons have charge proportional to mass and in a good approximation one has $f_c(DNA) = 1$ Hz independently of the length of the DNA strand. For $\hbar_{gr}(Fe^{++})$ $f_c(DNA)$ would correspond to $E = 1.86$ eV in the range of visible energies.

3.2 Trying to understand $\hbar_{eff}$ and $\hbar_{gr}$

Although $\hbar_{eff}$ and $\hbar_{gr}$ have become an essential part of quantum TGD, there are still many poorly understood aspects related to them.

3.2.1 Should one introduce a hierarchy of poly-local Planck constants?

The ordinary Planck constant is a universal constant and single-particle entity and serves as a quantization unit for local charges. $\hbar_{gr}$ depends on the masses of the members of the interacting systems and a bi-local character. This suggests that one should not mix these notions.

Both $\hbar_{gr}$ and its possible generalization to gauge interactions such as $\hbar_{em}$, would depend on the charges of the interacting particles. If they serve as charge units, the charges must be bilocal.

Should one introduce a hierarchy of poly-local Planck constants? Later a possible interpretation in terms of Yangian symmetries [1] [3, 2], which involve poly-local charges, will be considered. Each multi-local contribution to charge would involve its own Planck constant determined number theoretically.

Standard quantization rules for observables use $\hbar$ as a basic unit. Should one modify these rules by replacing $\hbar$ with (say) $\hbar_{em} = q_1 q_2 e^2/\alpha$ for $q_1 q_2 \alpha \geq 1$? Could these rules hold true at magnetic flux tubes characterized by $\hbar_{em}$? Could the charge units for the matter in the non-perturbative phase be $q_1 q_2$-multiples of the ordinary basic units? Could one find empirical evidence for the scaling up of the quantization unit in non-perturbative phases?

In order to avoid total confusion, one must distinguish clearly between the single particle Planck constant and its 2-particle and n-particle variants as Yangian picture suggests. One must also distinguish between p-adic CCE as a discrete counterpart of ordinary CCE and dark coupling constant evolution.

3.2.2 The counterpart of $\hbar_{gr}$ for gauge interactions

The gauge couplings g_i for various interactions disappear completely from the basic formulation of TGD since they are automatically absorbed into the definition of the induced gauge potentials. Hence $\beta_0/4\pi \equiv \alpha_K$ appears as a coupling parameter in the perturbative expansion based on the exponent of Kähler function. $\hbar$ or $\hbar_{eff}$ appear as charge unit only in the definition of conserved charges as Noether charges but not in the action exponential.

The generalization of the Nottale formula to other interactions is not quite obvious. Two-particle Planck constant $h_{eff}(2)$ is in question and GMm would be replaced with the product $q_1 q_2 g_i^2$. Since α_K determines all other coupling strengths so that it is enough to consider it.

The parameter $\beta_0/4\pi$ is analogous to fine structure constant since gravitational perturbative expansion is in powers of it [17].

β_0 is the gravitatoinal counterpart of the dimensionless coupling strength α_K defined in the QFT framework as a derived quantity $\alpha_K = g_K^2/4\pi$ but identified in the TGD context as the fundamental parameter appearing in Kähler action.

In TGD e does not appear as gauge coupling at the fundamental level (as opposed to QFT limit) but one can *define* e^2 as $e^2 = 4\pi\alpha\hbar$. α would obey p-adic CCE and $\hbar$ would be universal constant at single particle level. For dark phases, for which one has $h_{eff} > h$, $\alpha(1) \propto 2/n$, n dimension of the extension would hold true.

Consider the analog of the Nottale formula for em interactions. The coupling strength would be $q_1 q_2 e^2$ and for $q_1 q_2 e^2 \alpha > 1$, one would have

$$\hbar_{em}(2) = \frac{q_1 q_2 e^2}{\beta_0} \ . \tag{3.3}$$

This would give $\alpha_{em}(2) = \beta_0$. For $\beta_0 = \alpha$, one would obtain a coupling parameter α instead of $q_1 q_2 \alpha$ and the interpretation would be in terms of a transition to non-perturbative phase.

Does this phase transition correspond to a transition to dark phase? Could one interpret the phase transition by saying the dimension of extension is scaled by $n = \hbar_{em}(2)/\hbar$ identified as scaling of the dimension of extension of rationals?

Number theoretic vision predicts that in the dark evolution h_{eff} scales as n, the dimension of extension of rationals for all values of particle number in the definition of $h_{eff}(h)$ so that the single particle coupling constant strength would behave like $1/n$.

3.2.3   Charge fractionalization and the value of $\hbar_{eff}$

$\hbar_{eff} < \hbar$ implies charge fractionalization at the level of imbedding space. This inspires the question whether an analog of fractional quantum Hall effect could be in question. This is not the case.

1. The TGD based model for anyons [?] relies on the observation that the unit for the fractional quantization of transverse conductance in fractional quantum Hall effect (FQHE) as

$$\begin{aligned}
\sigma &= \nu \times \frac{e^2}{h} \ , \\
\nu &= \frac{n}{m} \ .
\end{aligned} \tag{3.4}$$

   The proposal is that FQHE could be understood as integer quantum Hall effect corresponding to $n \to kn$ for $h_{eff} = km\hbar$. $k = 1$ is the simplest possibility. Interestingly, the observed values of m are primes [13]: they would correspond to simple Galois groups Z_p in the TGD framework.

2. The fractionalization of charges could be understood at space-time level by noticing that n-sheetedness can be realized as analog of analytic function $z^{1/n}$. n full 2π turns are needed to return to the original point at space-time level so that it is possible to have fractional spin as multiples of $\hbar/n$. The many-particle states however have half-integer spin always since they correspond to representations of the Lorenz group as a symmetry group of $M^4 \times CP_2$. The action of rotations by multiples of 2π would correspond to the action of the Galois group.

   These two apparently conflicting mechanisms of charge fractionization correspond to two views about symmetries: either their act at the level of the embedding space or of space-time.

3. For $GMm/v_0 < \hbar$ one would have formally $\hbar_{eff} < \hbar$. Could this option make sense and give rise to a charge fractionalization? One can argue that for $\hbar_{gr} < \hbar_0$, $\hbar_0$ serves as the quantization unit and holds at the level of ordinary matter. This would give a condition $GMm \leq \beta_0$ to the product Mm of the masses involved.

   A stronger condition would hold true at single particle level and state $M/M_{Pl} \geq \sqrt{\beta_0}$ (or $M/M(CP_2) \geq \sqrt{\beta_0}$) for both masses involved. Dark quantum gravity would hold true only above Planck masses. In applications to elementary particle level this would require quantum coherent states of particles with total mass not smaller than Planck mass. Interestingly, a water blob with the size of a large cell has this size for $\beta_0 = 1/2$ [37].

3.2.4 What does the dependence of $\hbar_{gr}$ on particle masses mean?

$\hbar_{gr}$ depends on two masses. How could one interpret this geometrically?

1. The interpretation has been that a particle with energy E (and mass m) experiences the gravitational field of mass M via gravitational flux tubes characterized by $\hbar_{gr} = GME/v_0$ so that every particle has its specific gravitational flux tubes.

2. Could the thickness of the gravitational flux tubes correspond to the ordinary Compton length λ_c or gravitational Compton length $\lambda_{gr} = GM/v_0$? λ_c decreases with mass and λ_{gr} looks a more reasonable option concerning gravitational interaction.

3. At least static gravitational fields are analogous to static electric fields and in many-sheeted space-time the voltages as analogs of gravitational potential difference are the same along different space-time sheets. The same should hold for gravitational potential.

   Could one assume that gravitational potential has almost copies at all parallel sheets of the many-sheeted space-time (parallel with respect to M^4). Could these sheets correspond to different particle masses so that a particle with a given mass would have its own space-time sheet to represent its interactions with the central mass M.

4. These classical fields would be somehow represented by Kähler magnetic flux tubes carrying generalized Beltrami fields [4, 5, 6, 7] having also an electric part.

   Could these flux tubes somehow also represent the classical gravitational field? Could the electric part for the induced M^4 Kähler form predicted by the twistor lift of TGD [23, 25] giving rise to CP breaking, give a representation for the gravitational potential? Could this concretely realize the analogy between gravitation and electromagnetism?

5. A possible realization of this picture would be a fractal structure consisting of flux tubes within flux tubes emanating from the central mass. The radii of the flux tubes would decrease with m as long as $GMm/\beta > \hbar$ holds true. For smaller masses, the flux tube radius would correspond to Compton length.

   Fractal structures known as fractons (`https://en.wikipedia.org/wiki/Fracton`) are the recent hot topic of condensed matter physics (`https://cutt.ly/YQjqyjJ`. The explanation requires the replacement of the time evolution as a time translation with a scaling and condensed matter lattice would be replaced with fractal. These phases have exotic properties: in particular, thermal equilibrium need not be possible. There are also long range correlations due to fractality, which makes these phases ideal for quantum computation.

   In the TGD framework, the time evolutions between SSFRs are indeed generated by the scaling operator L_0 of super-conformal algebra and many-sheeted space-time is indeed both p-adic and dark fractal. The hierarchy of Planck constants makes possible quantum coherence in all scales.

3.2.5 Yangian symmetry and poly-local Planck constants

The product structure of $\hbar_{gr}$ and $\hbar_{em}$ has remained a mystery since it suggests that it characterizes the interaction of 2 space-time sheets whereas the ordinary Planck constant serves as a quantization unit for single particle states. Instead of a Galois group for a single space-time sheet, one would have a product of Galois groups for the two space-time sheets determined as roots for the polynomials in the product. Therefore one should write $\hbar_{gr} = \hbar_{eff,2}$ to distinguish it from a single particle Planck constant $\hbar_{eff}(1) \equiv \hbar_{eff}$.

1. In the TGD framework, wormhole contacts connecting two space-time sheets with Minkowskian signature are indeed building bricks of elementary particles and fundamental fermions appearing

as building bricks of elementary particles would be associated with the throats of the wormhole contact.

Could the two Minkowskian sheets be microscopically k-sheeted entities with sheets parallel to M^4 and perhaps determined as roots of a polynomial of degree k and having Galois group with order m? The maximal Galois group would be S_k with $m = k!$.

The scaling of $\hbar_0 \to \hbar(2)$ would mean that the pairs of these space-time surface sheets decompose to $\hbar(2)/\hbar_0$ pairs as orbit of $Gal \times Gal$ contributing to various quantum numbers a contribution proportional $\hbar_{gr}(2)/\hbar_0 = n_1 n_2 = k_1 k_2 m^2$.

The quantization unit would be $\hbar_0(2)$ for 2-particle quantities such as relative angular momentum. Spin is however thought to be single particle observable. The ordinary phase has a single-particle Planck constant as $\hbar(1)/\hbar_0 = m$.

2. There is no obvious reason for excluding the values of single particle $h_{eff}(1)/h_0$, which are considerably smaller than m or even equal to the minimal value $h_{eff}(1) = h_0$: they would correspond to Galois groups with smaller orders than $m = k!$ of say S_k.

   These exotic particles would have charge and spin units considerably smaller than $\hbar = m h_0$. Why have they not been observed (the findings of Mills are a possible exception and anyonic charge fractionization seems to be a different phenomenon)? Are these space-time sheets somehow unstable? Does gravitation somehow select the Galois group of stable ground state space-time surface so that R as a fundamental length scale is replaced with l_P as effective fundamental length?

3. Yangian algebras [1] [3, 2] involve besides single particle observables also $n > 1$-particle observables. Conserved charges have poly-local components which depend on n particles. Note that interaction energy represented as a potential energy is the simplest example about non-local 2-particle contribution to conserved energy.

   Yangian algebras are proposed to be central for TGD [29] and would reflect the replacement of the space-time locality with locality at the level of "world of classical worlds" (WCW) due to the replacement of a point like with a 3-surface, which can also consist of disjoint parts. Yangian picture suggests that single-particle $\hbar$ has n-particle generalization. The possible number theoretical rule could be

$$\frac{\hbar_{gr,n}}{\hbar_0} = \prod_k n_k \ , \tag{3.5}$$

where n_k correspond to the orders of Galois groups associated with the space-time sheets involved.

3.3 h/h_0 as the ratio of Planck mass and CP_2 mass?

Could one understand and perhaps even predict the value of h_0? Here number theory and the notion of n-particle Planck constant $h_{eff}(n)$ suggested by Yangian symmetry could serve as a guidelines.

1. Hitherto I have found no convincing empirical argument fixing the value of $r = \hbar/\hbar_0$: this is true for both single particle and 2-particle case.

   The value $h_0 = \hbar/6$ [27] as a maximal value of $\hbar_0$ is suggested by the findings of Randell Mills [12] and by the idea that spin and color must be representable as Galois symmetries so that the Galois group must contain $Z_6 = Z_2 \times Z_3$. Smaller values of h_0 cannot be however excluded.

2. A possible manner to understand the value r geometrically would be following. It has been assumed that CP_2 radius R defines a fundamental length scale in TGD and Planck length squared $l_P^2 =$

$\hbar G = x^{-2} \times 10^{-6} R^2$ defines a secondary length scale. For Planck mass squared one has $m^2_{Pl} = m(CP_2, \hbar)^2 \times 10^6 x^2$, $m(CP_2, \hbar)^2 = \hbar/R^2$. The estimate for x from p-adic mass calculations gives $x \simeq 4.2$. It is assumed that CP_2 length is fundamental and Planck length is a derived quantity.

But what if one assumes that Planck length identifiable as CP_2 radius is fundamental and CP_2 mass corresponds the minimal value h_0 of $h_{eff}(2)$? That the mass formula is quadratic and mass is assignable to wormhole contact connecting two space-time sheets suggests in the Yangian framework that $h_{eff}(2)$ is the correct Planck constant to consider.

One can indeed imagine an alternative interpretation. CP_2 length scale is deduced indirectly from p-adic mass calculation for electron mass assuming $h_{eff} = h$ and using Uncertainty Principle. This obviously leaves the possibility that $R = l_P$ apart from a numerical constant near unity, if the value of h_{eff} to be used in the mass calculations is actually $h_0 = (l_P/R)^2 \hbar$. This would fix the value of $\hbar_0$ uniquely.

The earlier interpretation makes sense if $R(CP_2)$ is interpreted as a dark length scale obtained scaling up l_P by $\hbar/\hbar_0$. Also the ordinary particles would be dark.

h_0 would be very small and $\alpha_K(\hbar_0) = (\hbar/\hbar_0)\alpha_K$ would be very large so that the perturbation theory for it would not converge. This would be the reason for why $\hbar$ and in some cases some smaller values of h_{eff} such as $\hbar/2$ and $\hbar/4$ [12] [27] seem to be realized.

For $R = l_P$ Nottale formula remains unchanged for the identification $M_P = \hbar/R$.

3.3.1 Various options

Number theoretical arguments allow to deduce precise value for the ratio $\hbar/\hbar_0$. Accepting the Yangian inspired picture, one can consider two options for what one means with $\hbar$.

1. $\hbar$ refers to the single particle Planck constant $\hbar_{eff}(1)$ natural for point-like particles.

2. $\hbar$ refers to $h_{eff}(2)$. This option is suggested by the proportionality $M^2 \propto \hbar$ in string models due to the proportionality $M^2 \propto \hbar/G$ in string models. At a deeper level, one has $M^2 \propto L_0$, where L_0 is a scaling generator and its spectrum has scale given by $\hbar$.

   Since M^2 is a p-adic thermal expectation of L_0 in the TGD framework, the situation is the same. This also due the fact that one has In TGD framework, the basic building bricks of particles are indeed pairs of wormhole throats.

One can consider two options for what happens in the scaling $h_{eff} \to k h_{eff}$.
Option 1: Masses are scaled by k and Compton lengths are unaffected.
Option 2: Compton lengths are scaled by k and masses are unaffected.
The interpretation of $M_P^2 = (\hbar/hbar_0)M^2(CP_2)$ assumes Option 1 whereas the new proposal would correspond to Option 2 actually assumed in various applications.

For Option 1 $m^2_{Pl} = (\hbar_{eff}/\hbar)M^2(CP_2)$. The value of $M^2(CP_2) = \hbar/R^2$ is deduced from the p-adic mass calculation for electron mass. One would have $R^2 \simeq (\hbar_{eff}/\hbar)l_P^2$ with $\hbar_{eff}/\hbar = 2.54 \times 10^7$. One could say that the real Planck length corresponds to R.

3.3.2 Quantum-classical correspondence favours Option 2)

In an attempt to select between these two options, one can take space-time picture as a guideline. The study of the imbeddings of the space-time surfaces with spherically symmetric metric carried out for almost 4 decades ago suggested that CP_2 radius R could naturally correspond to Planck length l_P. The argument is described in detail in Appendix and shows that the $l_P = R$ option with $h_{eff} = h$ used in the classical theory to determine α_K appearing in the mass formula is the most natural.

ISSN: 2153-8301   Prespacetime Journal   www.prespacetime.com
Published by QuantumDream, Inc.

3.3.3 Deduction of the value of $\hbar/\hbar_0$

Assuming Option 2), the questions are following.

1. Could $l_P = R$ be true apart from some numerical constant so that CP_2 mass $M(CP_2)$ would be given by $M(CP_2)^2 = \hbar_0/l_P^2$, where $\hbar_0 \simeq 2.4 \times 10^{-7}\hbar$ ($\hbar$ corresponds to $\hbar_{eff}(2)$) is the minimal value of $\hbar_{eff}(2)$. The value of $\hbar_0$ would be fixed by the requirement that classical theory is consistent with quantum theory! It will be assumed that $\hbar_0$ is also the minimal value of $\hbar_{eff}(1)$ both $\hbar_{eff}(2)$.

2. Could $\hbar(2)/\hbar_0(2) = n_0$ correspond to the order of the product of identical Galois groups for two Minkowskian space-time sheets connected by the wormhole contact serving as a building brick of elementary particles and be therefore be given as $n_0 = m^2$?

Assume that one has $n_0 = m^2$.

1. The natural assumption is that Galois symmetry of the ground state is maximal so that m corresponds to the order a maximal Galois group - that is permutation group S_k, where k is the degree of polynomial.

   This condition fixes the value k to $k = 7$ and gives $m = k! = 7! = 5040$ and gives $n_0 = (k!)^2 = 25401600 = 2.5401600 \times 10^7$. The value of $\hbar_0(2)/\hbar(2) = m^{-2}$ would be rather small as also the value of $\hbar_0(1)\hbar(1)$. p-Adic mass calculations lead to the estimate $m_{Pl}/m(CP_2) = \sqrt{m}m(CP_2) = 4.2 \times 10^3$, which is not far from $m = 5040$.

2. The interpretation of the product structure $S_7 \times S_7$ would be as a failure of irreducibility so that the polynomial decomposes into a product of polynomials - most naturally defined for causally isolated Minkowskian space-time sheets connected by a wormhole contact with Euclidian signature of metric representing a basic building brick of elementary particles.

   Each sheet would decompose to 7 sheets. $\hbar_{gr}$ would be 2-particle Planck constant $h_{eff}(2)$ to be distinguished from the ordinary Planck constant, which is single particle Planck constant and could be denoted by $h_{eff}(1)$.

   The normal subgroups of $S_7 \times S_7$ $S_7 \times A_7$ and $A_7 \times A_7$, S_7, A_7 and trivial group. A_7 is simple group and therefore does not have any normal subgroups expect the trivial one. S_7 and A_7 could be regarded as the Galois group of a single space-time sheet assignable to elementary particles. One can consider the possibility that in the gravitational sector all EQs are extensions of this extension so that $\hbar$ becomes effectively the unit of quantization and m_{Pl} the fundamental mass unit. Note however that for very small values of α_K in long p-adic length scales also the values of $h_{eff} < h$, even $\hbar_0$, are in principle possible.

   The large value of $\alpha_K \propto 1/\hbar_{eff}$ for Galois groups with order not considerably smaller than $m = (7!)^2$ suggests that very few values of $h_{eff}(2) < h$ are realized. Perhaps only $S_7 \times S_7$ $S_7 \times A_7$ and $A_7 \times A_7$ are allow by perturbation theory. Now however that in the "stringy phase" for which super-conformal invariance holds true, $\hbar_0$ might be realized as required by p-adic mass calculations. The alternative interpretation is that ordinary particles correspond to dark phase with R identified dark scale.

3. A_7 is the only normal subgroup of S_7 and also a simple group and one has $S_7/A_7 = Z_2$. $S_7 \times S^7$ has $S_7 \times S^7/A_7 \times A_7 = Z_2 \times Z_2$ with $n = n_0/4$ and $S_7 \times S^7/A_7 \times S_7 = Z_2$ with $n = n_0/2$. This would allow the values $\hbar/2$ and $\hbar/4$ as exotic values of Planck constant.

   The atomic energy levels scale like $1/\hbar^2$ and would be scaled up by factor 4 or 16 for these two options. It is not clear whether $\hbar \to \hbar/2$ option can explain all findings of Randel Mills [12] in TGD framework [27], which effectively scale down the principal quantum number n from n to $n/2$.

ISSN: 2153-8301   Prespacetime Journal   www.prespacetime.com
                              Published by QuantumDream, Inc.

4. The product structure of the Nottale formula suggests

$$n = n_1 \times n_2 = k_1 k_2 m^2 \; . \tag{3.6}$$

Equivalently, n_i would be a multiple of m. One could say that $M_{Pl} = \sqrt{\hbar/\hbar_0} M(CP_2)$ effectively replaces $M(CP_2)$ as a mass unit. At the level of polynomials this would mean that polynomials are composites $P \circ P_0$ where P_0 is ground state polynomial and has a Galois group with degree n_0. Perhaps S_7 could be called the gravitational or ground state Galois group.

3.4 Connection with adelic physics and infinite primes

The structure of $\hbar_{gr}$ and its electromagnetic counterpart $\hbar_{em}$ characterize 2-particle states whereas $\hbar$ characterizes single particle state. Yangian picture suggests that the notion of $\hbar_{eff}(n)$, $n = 1, 2, ..$ makes sense.

One can decompose a state consisting of N particles in several manners to partitions consisting of m subsets with n_i, $i = 1, ..., n$ in a given subset of particles. Could these subsets correspond to gravitationally bound states so that one can take these sets as basic entities characterized by masses and assume that gravitational interactions reduce to gravitational interactions between them and are quantal for $GM_i M_j / v_0 \geq \hbar$. Same question applied to electromagnetic, weak and color interactions.

3.4.1 Connection with adelic physics

This picture would have analog at the level of adelic physics [38, 39, 44].

1. In the M^8 picture space-time surfaces correspond to "roots" of complexified octonionic polynomials obtained from irreducible real polynomials with rational (or perhaps even algebraic) coefficients. The dynamics realizes associativity of the normal space of the complexified space-time surface having 4-D space-time surface as real part mapped from M^8 to $H = M^4 \times CP_2$ by $M^8 - H$ correspondence.

2. One can consider irreducible polynomials of several variables such that the additional variables are interpreted as parameters [42]. The parametrized set of polynomials defines a parametrized set of space-time surfaces and one can have a superposition of quantum states corresponding to irreducible polynomial of degree n and products of irreducible polynomials with sum of degrees n_i equal to n. This kind of parametrized set could define sub-spaces of the "world of classical worlds" (WCW).

3. Irreducibility fails for some parameter values forming lower-dimensional manifolds of the parameter space. The failure of the irreducibility means decomposition to a product of polynomials in which the set of roots decomposes to subsets R_i, which are roots of a rational polynomial with a lower degree n_i. Spacetime surface as a coherent structure decomposes to uncorrelated space-time surfaces with a discrete set of points as intersections. In this manner one obtains a decomposition of the parameter space to subsets of decreasing dimension. The generic situation has maximal dimension and dimension equal to that of the parameter space.

4. The catastrophe theory [10] founded by Rene Thom studies these situations. In catastrophe theory, the failure of the irreducibility is of very special nature and means that some roots of the polynomial co-incide and become multiple roots. For polynomials with rational coefficients, they would become multiple rational roots so that the degree of the polynomial determining the extension would be reduced by two units. This is discussed in detail from TGD point of view in [42]. For polynomials with rational coefficients, typically complex conjugate roots become rational and the dimension of the algebraic extension is reduced.

ISSN: 2153-8301   Prespacetime Journal   www.prespacetime.com
Published by QuantumDream, Inc.

5. The quantum state defined by the polynomial of several variables would be a superposition of space-time surfaces labelled by the points of the parameter space. It would decompose to subsets defining what is known as a stratification. The subsets for which the polynomial fails to be irreducible would have lower dimension. For polynomials with rational coefficients these sets would be discrete and it is not clear whether the lower-dimensional sets are non-empty in the generic case.

6. The decomposition to k irreducible polynomials with degrees n_i, $i = 1, ..., k$ would correspond to a decomposition of the space-time surface to separate space-time surfaces with $h_{gr,i} = n_i h_0 = GM_i m/v_0$ (same applies to h_{em}) satisfying $\sum n_i = n$. These would correspond to different decompositions of the total energy to a sum of energies E_i: $E = \sum E_i$. The irreducible polynomials with degree n_i could be interpreted as bound states for a subset of basic units. Maximal decomposition would correspond to $n_i = 1$ and have interpretation as a set of elementary particles with $h_{eff} = h_0$ (note that $h = 6h_0$ in the proposal inspired by the findings of Randel Mills [27]).

3.4.2 Connection with infinite primes

The notion of infinite prime [24] resonates with this picture.

1. The hierarchy of infinite primes has an interpretation as a repeated second quantization of supersymmetric arithmetic QFT. Polynomial primes of variable polynomials of single variable with rational coefficients follow ordinary primes in the hierarchy. Higher levels correspond to polynomial primes for polynomials of several variables and second quantization corresponds to the formation of polynomials of single variable with coefficients as polynomials of $n - 1$ variables.

   Irreducible polynomials of higher than first order have interpretation as bound states whereas polynomials reducing to products of monomials correspond to Fock states of free particles.

2. The beatiful feature would be a number theoretic description of also bound states. The description of the particle decays as a failure of the irreducibility of the polynomials corresponding to infinite primes would extend this picture to the dynamics.

3. Second beautiful feature is the number theoretic description of particle reactions. Particle reactions with unentagled final states would naturally correspond to a situation in which the initial (prepared) and final (state function reduced) states are products of polynomials. Interaction period would correspond to an irreducible polynomial.

   This picture conforms with the proposal inspired originally by a model of "cold fusion". unnelling phenomenon crucial for nuclear reactions would correspond to a formation of dark phase in which the value of h_{eff} increases [35, 28, 40]. This picture generalizes to all particle reactions.

4 How to understand coupling constant evolution?

In this section, the evolutions of Kähler coupling strength α_K and gravitational fine structure constant α_{gr} are discussed. The reason for restricting to α_K is that it is expected to induce the evolution of various gauge couplings, and could also induce the evolution of α_{gr}.

4.1 Evolution of Kähler coupling strength

The evolution of Kähler coupling strength $\alpha_K = g_K^2/2h_{eff}$ gives the evolution of α_K as a function of dimension n of EQ: $\alpha_K = g_K^2/2nh_0$. If g_K^2 corresponds to electroweak U(1) coupling, it is expected to evolve also with respect to PLS so that the evolutions would factorize.

Note that the original proposal that g_K^2 is renormalization group invariant was later replaced with a piecewise constancy: α_K has indeed interpretation as piecewise constant critical temperature

1. In the TGD framework, coupling constant as a continuous function of the continuous length scale is replaced with a function of PLS so that coupling constant is a piecewise constant function of the continuous length scale.

   PLSs correspond to p-adic primes p, and a hitherto unanswered question is whether the extension determines p and whether p-adic primes possible for a given extension could correspond to ramified primes of the extension appearing as factors of the moduli square for the differences of the roots defining the space-time surface.

   In the M^8 picture the moduli squared for differences $r_i - r_j$ of the roots of the real polynomial with rational coefficients associated with the space-time surfaces correspond to energy squared and mass squared. This is the case of p-adic prime corresponds to the size scale of the CD.

   The scaling of the roots by constant factor however leaves the number theoretic properties of the extension unaffected, which suggests that PLS evolution and dark evolution factorize in the sense that PLS reduces to the evolution of a power of a scaling factor multiplying all roots.

2. If the exponent $\Delta K/log(p)$ appearing in $p^{\Delta K/log(p))} = exp(\Delta K)$ is an integer, $exp(\Delta K)$ reduces to an integer power of p and exists p-adically. If ΔK corresponds to a deviation from the Kähler function of WCW for a particular path in the tree inside CD, p is fixed and $exp(\Delta K)$ is integer. This would provide the long-sought-for identification of the preferred p-adic prime. Note that p must be same for all paths of the tree. p need not be a ramified prime so that the trouble-some correlation between n and ramified prime defining padic prime p is not required.

3. This picture makes it possible to understand also PLS evolution if ΔK is identified as a deviation from the Kähler function. $p^{\Delta K/log(p))} = exp(\Delta K)$ implies that ΔK is proportional to $log(p)$. Since ΔK as 6-D Kähler action is proportional to $1/\alpha_K$, $log(p)$-proportionality of ΔK could be interpreted as a logarithmic renormalization factor of $\alpha_K \propto 1/log(p)$.

4. The universal CCE for α_K inside CDs would induce other CCEs, perhaps according to the scenario based on Möbius transformations [32].

4.1.1 Dark and p-adic length scale evolutions of Kähler coupling strength

The original hypothesis for dark CCE was that $h_{eff} = nh$ is satisfied. Here n would be the dimension of EQ defined by the polynomial defining the space-time surface $X^4 \subset M_c^8$ mapped to H by $M^8 - H$ correspondence. n would also define the order of the Galois group and in general larger than the degree of the irreducible polynomial.

Remark: The number of roots of the extension is in general smaller and equal to n for cyclic extensions only. Therefore the number of sheets of the complexified space-time surface in M_c^8 as the number of roots identifiable as the degree d of the irreducible polynomial would in general be smaller than n. n would be equal to the number of roots only for cyclic extensions (unfortunately, some former articles contain the obviously wrong statement $d = n$).

Later the findings of Randell Mills [12], suggesting that h is not a minimal value of h_{eff}, forced to consider the formula $h_{eff} = nh_0$, $h_0 = h/6$, as the simplest formula consistent with the findings of Mills [27]. h_0 could however be a multiple of even smaller value of h_{eff}, call if h_0 and the formula $h_0 = h/6$ could be replaced by an approximate formula.

The value of $h_{eff} = nh_0$ can be understood by noticing that Galois symmetry permutes "fundamental regions" of the space-time surface so that action is n times the action for this kind of region. Effectively this means the replacement of α_K with α_K/n and implies the convergence of the perturbation theory. This was actually one of the basic physical motivations for the hierarchy of Planck constants. In the previous section, it was argued that $\hbar/h_0$ is given by the ratio R^2/l_P^2 with R identified as dark scale equals to $n_0 = (7!)^2$.

ISSN: 2153-8301   Prespacetime Journal   www.prespacetime.com
Published by QuantumDream, Inc.

The basic challenge is to understand p-adic length scale evolutions of the basic gauge couplings. The coupling strengths should have a roughly logarithmic dependence on the p-adic length scale $p \simeq 2^{k/2}$ and this provides a strong number theoretic constraint in the adelic physics framework.

Since Kähler coupling strength α_K induces the other CCEs it is enough to consider the evolution of α_K.

4.1.2 p-Adic CCE of α from its value at atomic length scale?

If one combines the observation that fine structure constant is rather near to the inverse of the prime $p = 137$ with PLS, one ends up with a number theoretic idea leading to a formula for α_K as a function of p-adic length scale.

1. The fine structure constant in atomic length scale $L(k = 137)$ is given $\alpha(k) = e^2/2h \simeq 1/137$. This finding has created a lot of speculative numerology.

2. The PLS $L(k) = 2^{k/2}R(CP_2)$ assignable to atomic length scale $p \simeq 2^k$ corresponds to $k = 137$ and in this scale α is rather near to $1/137$. The notion of fine structure constant emerged in atomic physics. Is this just an accident, cosmic joke, or does this tell something very deep about CCE?

   Could the formula

   $$\alpha(k) = \frac{e^2(k)}{2h} = \frac{1}{k}$$

   hold true?

There are obvious objections against the proposal.

1. α is length scale dependent and the formula in the electron length scale is only approximate. In the weak boson scale one has $\alpha \simeq 1/127$ rather than $\alpha = 1/89$.

2. There are also other interactions and one can assign to them coupling constant strengths. Why electromagnetic interactions in electron Compton scale or atomic length scales would be so special?

The idea is however plausible since beta functions satisfy first order differential equation with respect to the scale parameter so that single value of coupling strength determines the entire evolution.

4.1.3 p-Adic CCE from the condition $\alpha_K(k = 137) = 1/137$

In the TGD framework, Kähler coupling strength α_K serves as the fundamental coupling strength. All other coupling strengths are expressible in terms of α_K, and in [32] it is proposed that Möbius transformations relate other coupling strengths to α_K. If α_K is identified as electroweak $U(1)$ coupling strength, its value in atomic scale $L(k = 137)$ cannot be far from $1/137$.

The factorization of dark and p-adic CCEs means that the effective Planck constant $h_{eff}(n, h, p)$ satisfies

$$h_{eff}(n, h, p) = h_{eff}(n, h) = nh \ . \tag{4.1}$$

and is independent of the p-adic length scale. Here n would be the dimension of the extension of rationals involved. $h_{eff}(1, h, p)$ corresponding to trivial extension would correspond to the p-adic CCE as the TGD counterpart of the ordinary evolution.

The value of h need not be the minimal one as already the findings of Randel Mills [12] suggest so that one would have $h = n_0 h_0$.

$$h_{eff} = nn_0 h \ , \quad \alpha_{K,0} = \frac{g^2_{K,max}}{2h_0} = n_0 \ . \tag{4.2}$$

This would mean that the ordinary coupling constant would be associated with the non-trivial extension of rationals.

Consider now this picture in more detail.

1. Since dark and p-adic length scale evolutions factorize, one has

$$\alpha_K(n) = \frac{g^2_K(k)}{2h_{eff}} \ , \quad h_{eff} = nh_0 \ . \tag{4.3}$$

$U(1)$ coupling indeed evolves with the p-adic length scale, and if one assumes that $g^2_K(k, n_0)$ ($h = n_0 h_0$) is inversely proportional to the logarithm of p-adic length scale, one obtains

$$\begin{aligned} g^2_K(k, n_0) &= \frac{g^2_K(max)}{k} \ , \\ \alpha_K &= \frac{g^2_K(max)}{2kh_{eff}} \ . \end{aligned} \tag{4.4}$$

2. Since $k = 137$ is prime (here number theoretical physics shows its power!), the condition $\alpha_K(k = 137, h_0) = 1/137$ gives

$$\frac{g^2_K(max)}{2h_0} = \alpha_K(max) = (7!)^2 \ . \tag{4.5}$$

The number theoretical miracle would fix the value of $\alpha_K(max)$ to the ratio of Planck mass and CP_2 mass $n_0 = M_P^2/M^2(CP_2) = (7!)^2$ if one takes the argument of the previous section seriously.

The convergence of perturbation theory could be possible also for $h_{eff} = h_0$ if the p-adic length scale $L(k)$ is long enough to make $\alpha_K = n_0/k$ small enough.

3. The outcome is a very simple formula for α_K

$$\alpha_K(n, k) = \frac{n_0}{kn} \ , \tag{4.6}$$

$$\tag{4.7}$$

which is a testable prediction if one assumes that it corresponds to electroweak $U(1)$ coupling strength at QFT limit of TGD. This formula would give a practically vanishing value of α_K for very large values of n associated with h_{gr}. Here one must have $n > n_0$.

For $h_{eff} = nn_0 h$ characterizing extensions of extension with $h_{eff} = h$ one can write

$$\alpha_K(nn_0, k) = \frac{1}{kn} \ . \tag{4.8}$$

4. The almost vanishing of α_K for the very large values of n associated with $\hbar_{gr}$ would practically eliminate the gauge interactions of the dark matter at gravitational flux tubes but leave gravitational interactions, whose coupling strength would be $\beta_0/4\pi$. The dark matter at gravitational flux tubes would be highly analogous to ordinary dark matter.

ISSN: 2153-8301   Prespacetime Journal   www.prespacetime.com
Published by QuantumDream, Inc.

4.2   The evolution of the gravitational fine structure constant

Nottale [11] introduced the notion of gravitational Planck constant $\hbar_{gr} = GMm/\beta_0$ ($\beta_0 = v_0/c$ is velocity parameter), which has gigantic values so that the original proposal $\hbar_{gr} = nh_0$ would predict very large values for n. If p-adic and dark evolutions are independent this is not a problem since p-adic length scales need not be gigantic.

4.2.1   Evolution of the parameter β_0

Gravitational fine structure constant is given by $\alpha_{gr} = GMm/4\pi\hbar_{gr} = \beta_0/4\pi$. The basic challenge is to understand the value spectrum of β_0.

1. Kepler's law $\beta^2 = GM/r = r_S/2r$ suggests length scale evolution of form

$$\beta_N = \sqrt{\frac{r_S}{2L(N)x}} = \frac{\beta_{0,max}}{N} \quad . \tag{4.9}$$

   The coefficient x has been included in the formula because otherwise a conflict with Bohr model for planetary orbits results.

2. How to identify N?

   (a) $N = n = h_{gr}/h_0$ would give a gigantic value of N and this would give extremely small value for β_0. Actually $N = n$ for n in $h_{gr} = nh_0$ is impossible as is clear from the defining equation.

   (b) It is not clear whether N be identified as a dimension for some factor in the composition of extension to simple factors rather than as n. This would conform with the vision that there are evolutionary hierarchies of extensions of extensions of... for which the dimension is product of dimensions of the extensions involved.

   (c) The simplest option is that p-adic length scale evolution determines N as in case of the gauge interactions, and it corresponds to k in $p \simeq 2^k$. $log_2(p)$ exists also for a general prime p in real sense. In p-adic sense it exists for all primes except $p = 2$ as integer valued function. $p = 2$ could be chosen to be the exceptional prime.

   This would conform with the idea that gravitational sector and gauge interaction sector correspond to different factors in the decomposition of extension of rationals. Perhaps the gravitational part of EQ extends its gauge part. This would conform with the idea that gravitation does not differentiate between states with different gauge quantum numbers.

What can one say about the value of $\beta_{0,max}$ and its length scale evolution?

1. The value of $\beta_{0,max} = 1/2$ would give for the length scale $L = GM/\beta_{0,max} = r_S$. If one requires that the scale L is not smaller than Scwartschild radius, $\beta_{0,max} \leq 1/2$ follows. $\beta_{0,max} = 1/2$ is the first guess but it turns that number theoretical constraintss exclude it and suggest $\beta_{0,max} = \pi/6$ as the simplest guess.

2. Gravitational Bohr radius a_{gr} given by

$$a_{gr} = \frac{\hbar_{gr}}{\alpha_{gr}m} per. \tag{4.10}$$

   defines a good candidate for the minimal value of L_n as $L_1 = a_{gr}$.

ISSN: 2153-8301   Prespacetime Journal   www.prespacetime.com
Published by QuantumDream, Inc.

3. The analogs of p-adic length scales would be equal to the radii of gravitational Bohr atom as n^2-multiples of the gravitational Bohr radius a_{gr}:

$$L_n = n^2 a_{gr} \quad , \quad a_{gr} = \frac{4\pi GM}{\beta_0^2} \quad . \tag{4.11}$$

This expression realizes the condition $\beta_0^2 = xGM/r$ inspired by the Kepler's law with $x = 4\pi$.

4. One must fix a_{gr} as a multiple $a_{gr} = kr_S$ of r_S. Substitution to the above equation gives

$$\beta_{0,max} = \sqrt{\frac{2\pi}{k}} \quad .$$

The condition $\beta_{0,max} = 1/2$ would give $k = 8\pi$ and $a_{gr} = 8\pi r_S$ as a minimal radius for a Bohr orbit. The condition $\beta_{0,max} < 1$ gives $k \geq 2\pi$ and $a_{gr} \geq 2\pi r_S$.

Just as in the case of hydrogen atom, the falling of the orbiting system to the blackhole like entity (in TGD frameworkd blackholes are replaced with what might be called flux tube spaghettis [34, 33]) is prevented. This should have obviously consequences for the view about the dynamics around blackhole like objects. The circular orbits have as analogs s-waves and of these are realized, the falling to blackhole like entity is possible.

5. The proposed formula does not force the condition $\beta_0 < 1$ and it is not clear whether it holds true at the relativistic limit. The replacement $\beta_0 \to sinh(\eta) = \beta_0/\sqrt{1 - \beta_0^2}$, where η is the hyperbolic angle, forces the condition $\beta_0 < 1$, and would give

$$\beta_0 \to \frac{\beta_0}{\sqrt{1 - \beta_0^2}} = \sqrt{\frac{2\pi}{k}} \quad .$$

The condition $\beta_{0,max} = 1/2$ gives $k/2\pi = 3$. This would correspond to the minimal Bohr radius $a_{gr} = 6\pi r_S \simeq 18.84 r_S$.

4.2.2 Number theoretical universality as a constraint

Also number theoretical universality could be also used as a constraint. The condition would be that only finite-dimensional extensions are allowed. π defines an infinite-D transcendental extension so that it should disappear in central formulas.

1. The appearance of 4π in the formula $a_{gr} = 4\pi G/\beta_{0,max}^2$ creates number-theoretical worries. Suppose that a_{gr} is a rational number.

2. I have proposed that G is dynamically determined and relates to the CP_2 radius via the formula $G = R^2/\hbar_{grav} = 2\pi R^2/\hbar_{grav}$, where $h_{grav}/h_0 \sim 10^7$ holds true [19].

   This gives

$$a_{gr} = \frac{4\pi G}{\beta_{0,max}^2} = \frac{8\pi^2 R^2}{h_{grav}\beta_{0,max}^2} \tag{4.12}$$

3. Since $\beta_0/4\pi$ appears as coupling strength in the perturbation theory, it should also be rational. $\beta_{0,max} = \pi/6$ would realize the condition $\beta_{0,max} = 1/2$ approximately.

4. With this assumption the rationality of a_{gr} requires that h_{gr} is proportional to π so that also G would be rational. This implies that $\hbar_{eff} = h_{eff}/2\pi$ is rational. Also α_K would be rational if g_K^2 is rational. This would be true also for the other coupling constants.

5. $\beta_0 = \pi/6$ would realize the condition $\beta_0 = 1/2$ approximately. This also implies that α_{gr} is rational. The condition $k/2\pi = 1/\beta_{0,max}^2$ implies $k \propto 1/\pi$. $a_{gr} = kr_s = kGM$ is rational, and this requires $M \propto \pi$. This guarantees the rationality of GM/β_0. Gravitational fine structure constant α_{gr} would be an inverse integer multiple of $\alpha_{gr}(max) = 1/24$. It would seem that the system is consistent.

   The alternative condition $\beta_0^2/(1 - \beta_0^2) = 2\pi/k$ is excluded because it implies that k is a rather complex transcendental.

What makes this interesting is that 24 is one of the magic numbers of mathematics (`https://cutt.ly/RnOxOTr`) and it appears in the bosonic string model as the number of space-like dimensions.

1. Euclidian string world sheet with torus topology has a conformal equivalence class defined by the ratio ω_2/ω_1 of the complex vectors spanning the parallelogram defining torus as an analog of a unit cell. String theory must be invariant under modular group $SL(2, Z)$ leaving the periods and thus the conformal equivalence class of torus invariant. Same applies to higher genera. In TGD these surfaces correspond to partonic 2-surfaces.

2. Modular invariance raises elliptic functions (doubly periodic analytic functions in complex plane) in a special role. In particular, Weierstrass function, which satisfies the differential equation $(d\mathcal{P}/dz)^2 = 4\mathcal{P}^3 - g_2\mathcal{P} - g_3$ has a key role in the theory of elliptic functions (`https://cutt.ly/BnOxrMS`).

   The discriminant $\Delta = g_2^3 - 3g_2^3$ of the polynomial at the r.h.s can be locally regarded as a function of the ratio of $\tau = \omega_2/\omega_1$ of the periods of $\mathcal{P}$ defining the conformal equivalence class of torus.

   $\Delta(\tau)$ is not a genuine modular invariant function of τ. Rather, Δ defines a modular form of weight 12 transforming as $\Delta(a\tau + b/(c\tau + d)) \to (c\tau + d)^{12}\Delta(\tau)$ under $SL(2, Z)$. The number 24 comes from the fact that one can express Δ as 24th power of the Dedekind η function: $\Delta = (2\pi)^{12}\eta^{24}$.

3. In dimension $D = 24$ there are 24 even positive definite unimodular lattices, called the Niemeier lattices, and the so-called Leech lattice is one of them. Interestingly, in dimension 4 there exists a 24-cell analogous to Platonic solid having 24 octahedrons as its 3-D "faces".

This encourages the question whether there might be a connection between TGD and string theory based views of quantum gravitation.

4.2.3 Test cases for the proposal

Phase transitions changing β_0 are possible at $r_n/a_{gr} = n^2$ at the Bohr orbits. For instance, in the Bohr orbit model the orbit of Earth is such an orbit. It can be regarded as $n = 5$ orbital with $\beta_0 \simeq 2^{-11}$ and is nearly circular so that the phase transition with $n = 1$ orbital with $\beta_0 \to \beta_0/5$ is possible. The outer planets indeed have $\beta_0/5$.

p-Adic length scale hierarchy is replaced union of hierarchies with $\beta_0 = \beta_{0,max}/n = 1/2n$, each of which is a subset of the set of Bohr orbits for $\beta_0 = \beta_{0,max}$. One can test this hypothesis for the proposed applications [47].

1. In the Bohr orbit model the inner planets Mercury, Venus, and Earth identifiable correspond to $n = 3, 4, 5$ orbitals for $\beta_0 \simeq 2^{-11}$. Solar radius is $R_{Sun} \simeq .7$ Gm. The orbital radius of Mercury is $R_M \simeq 58$ Gm $= 82.9 \times R_{Sun}$. This gives $a_{gr} = R_M/9 \simeq 9.2R_{sun}$. This gives $\beta_0 = \sqrt{2\pi R_S/a_{gr}} \simeq 17.1 * 10^{-4}$.

   The approximation used hitherto has been $\beta_0 = 2^{-11)} \simeq 5 \times 10^{-4}$ and is by a factor about $1/3$ smaller. Using $a_{gr} = R_M$ instead of $a_{gr} = R_M/9$ would give roughly correct value.

One could indeed regard Mercury as $n = 1$ orbit for $v_0 = v_0/3$ in which case one would have $a_{gr} = R_M$ and one would obtain $\beta_0 = .57$ which is not far from the valued used. Mercury would therefore correspond to $n = 3$ dark matter gravitationally whereas Venus must correspond to $n = 1, 2$ or $n = 4$.

2. The transition $\beta_0 \to \beta_0/5$ possible for Earth and required for outer planets could be interpreted as the increase of n having interpretation as increase of dimension of extension of rationals $n \to 5n$.

For the Earth one has $R_E = 6.371 \times 10^6$ m and $r_S = 10^{-2}$ m. The model of the superfluid fountain effect [20][47] suggests $\beta_0 = 1/2$ for which one would have $GM/v_0 = 1/2$. The value of $a_{gr} = 6\pi r_S$ for the relativistic form of the Nottale condition. The principal quantum number n for the Bohr orbit of the super-fluid would be $n \simeq R_E/a_{gr} = R_E/6\pi r_S \simeq 3.4 \times 10^7$. This would correspond to the large quantum number limit. The difference of radii between nearby Bohr orbits would be $\Delta r = 2R_E/n \simeq 19$ cm, which makes sense.

The levels in the hierarchy of gravitationally dark matters are labelled by $h_{gr} = GMm/\beta_0$ with $\beta_0 = \beta_{0,max}/n$, where n is the dimension of EQ, and each level defines a hierarchy of atomic orbitals. The sets of orbital radii at various levels form a nested hierarchy and phase transitions can occur at least between the states with the same angular momentum and orbital radius.

The quantum variant of the similar picture is expected to apply in the case of the hydrogen atom and the fact that there is evidence for dark valence electrons suggests that these phase transitions indeed take place.

What about long cosmic strings thickened to flux tubes explaining galactic dark matter in the TGD framework? In this case the Kepler law gives $\beta^2 = TG$ so that the all orbiting stars would correspond to the same value of β_0 and n.

5 Appendix: Imbedding of spherically symmetric stationary symmetric metric as a guideline

There are two basic questions to be answered.

1. Is $R = l_P$ or $R = m^2 l_P$, $m = 7!$ realized?

2. Should one assume that $g_K^2 \propto \hbar_{eff}$ or $\alpha_K \propto 1/\hbar_{eff}$?

For the first option α_K is the same for dark phases but would be subject to p-adic CCE. This would conform with the notion of gravitational Planck constant predicting that the parameter. The *effective* value of α_K would be however given by α_K/n for dark phases since the Galois symmetry is n-fold multiple of the action for a "fundamental region" for the Galois group.

Second option would predict that α_K behaves like $1/n$ so that effective α_K would behave like $1/n^2$. It seems that this option is excluded and one can concentrate on the first question. The increase of g_K^2 with n is not a problem since it does not appear as a parameter of perturbative expansion since g_K is automatically absorbed to a scaling of the induced gauge potentials.

Quantum-classical correspondence suggests that classical theory theory, in particular spherically symmetric stationary imbeddings, could help to answer the first question. Even the extremal property is not absolutely necessary.

The action is a sum of Kähler action and volume term proportional to length scale dependent cosmological constant approaching zero in long length scale and in equilibrium both give contributions of the same order of magnitude. This suggests that Kähler action corresponding to $\Lambda = 0$ could serve as a guideline.

I studied the embedding of a stationary spherically symmetric metric as a space-time surface during the first 10 years of TGD and the results suggested that the $R = l_P$ option looks more realistic. p-Adic

ISSN: 2153-8301   Prespacetime Journal   www.prespacetime.com
Published by QuantumDream, Inc.

mass calculations based on the definition of the Compton length as $\hbar/M$ however led to the conclusion that the one must have $r \sim 10^{7.6} l_P$. If one replaces $\hbar$ with $\hbar_0$, $R = l_P$ is natural.

The spherically symmetric ansatz assumes that space-time surfaces has a projection to a geodesic sphere S^2 of CP_2 which can be either homologically trivial or non-trivial. Using spherical coordinates (Θ, Φ) for S^2 and spherical Minkowski (t, r, θ, ϕ) coordinates for M^4, the ansatz reads

$$s \equiv sin(\Theta) = f(r) \ , \quad \Phi = \omega t \ ,$$

$$g_{tt} = 1 - k^2 s^2 \ , \qquad k^2 = R^2 \omega^2 \ . \tag{5.1}$$

In far-away region one can approximate s as

$$s = s_0 + \tfrac{r_1}{r}, \quad s_0 = sin(\Theta_0) \ . \tag{5.2}$$

The induced metric has component g_{tt} given by

$$g_{tt} = 1 - k^2 s_0^2 - 2k^2 s_0 \frac{r_1}{r} \ , \tag{5.3}$$

by taking $u = t\sqrt{2 - k^2 s_0^2}$ as a new time coordinate can expresses g_{tt} in terms of the parameters of Scwartshild metric

$$g_{uu} = 1 - 2k^2 s_0 \tfrac{r_1}{r} \equiv 1 - \tfrac{r_s}{r} \ ,$$

$$r_s = 2GM = \tfrac{2k^2 s_0 r_1}{1 - k^2 s_0^2} \ , \tag{5.4}$$

$$r_1 = \tfrac{1 - k^2 s_0^2}{2k^2 s_0} r_s \equiv k_1 r_s \ .$$

The approximation makes sense for $s \leq 1$, which gives the condition

$$r \geq r_{min} = (1 - s_0)r_1 = (1 - s_0)k_1 r_s = (1 - s_0)\frac{1 - k^2 s_0^2}{2k^2 s_0} r_s \equiv y_1 r_s \ . \tag{5.5}$$

Remark: The radial component of the metric goes to zero much faster than for Schwartschild metric. The shift of time coordinates depending on the radial coordinate allows to correct this problem. This is however not essential for the recent argument. Schwartchild metric however implies that $\sqrt{g}$ in the calculation of mass gives just the volume element of the flat metric since $g_{tt} g_{rr} = 1$ is true. This is assumed in the following.

One can estimate the mass of the system as Kähler electric energy. Assume that the contribution to the mass comes only from the region $r > y_1 r_s$. The Kähler electric mass $M = r_s/2G$ is given by the expression

$$M = \tfrac{r_s}{2G}$$

$$= \tfrac{\hbar_{eff}}{2\alpha_K} \tfrac{s_0^2}{1 - s_0^2} r_1^2 \omega^2 \int_{r_{min}}^{\infty} \tfrac{dr}{r^2} = \tfrac{\hbar_{eff}}{2\alpha_K} \tfrac{(1 - k^2 s_0^2)s_0}{2(1 - s_0)} r_s \tfrac{1}{R^2} \ . \tag{5.6}$$

This gives a consistency condition relating R and l_P

$$R^2 = \frac{\hbar_{eff}}{\hbar} X l_P^2 \;,$$

$$X = \frac{(1 - k^2 s_0^2) s_0}{\alpha_K (1 - s_0)} \;. \tag{5.7}$$

One can consider two cases.

1. For $\hbar_{eff} = \hbar$ the condition reduces to

$$R^2 = X l_P^2 \;. \tag{5.8}$$

$l_P = R$ gives $X = (1 - k^2 s_0^2) s_0 / \alpha_K (1 - s_0) = 1$. One should have $s_0 \simeq \alpha_K$ so that the value of $1/\alpha_K$ as an analog of critical temperature would be coded to the geometry of the space-time surface. $R = (7!)^2 l_P$ would require $X = \hbar/\hbar_0$, one should have $1 - s_0 \sim 10^{-5}$ for $\alpha_K \sim 10^{-2}$.

2. For $\hbar_{eff} = \hbar_0$ the condition reduces to

$$R^2 = X \frac{\hbar_0}{\hbar} \times l_P^2 \;. \tag{5.9}$$

$l_P = R$ gives $X = \hbar/\hbar_0$. One might of course argue that α_K decreases in long scales in the discrete p-adic length scale evolution but this option does not look plausible.

To sum up, intuitively $\hbar$ option with $R = l_P$ looks the most reasonable option.

Received June 28, 2021; Revised August 4, 2021; Accepted December 4, 2021

References

[1] Yangian symmetry. Available at: `http://en.wikipedia.org/wiki/Yangian`.

[2] Witten E Dolan L, Nappi CR. Yangian Symmetry in $D = 4$ superconformal Yang-Mills theory, 2004. Available at: `http://arxiv.org/abs/hep-th/0401243`.

[3] Loebbert F. Lectures on Yangian Symmetry, 2016. Available at: `https://arxiv.org/pdf/1606.02947.pdf`.

[4] Lakthakia A. *Beltrami Fields in Chiral Media*, volume 2. World Scientific, Singapore, 1994.

[5] Bogoyavlenskij OI. Exact unsteady solutions to the Navier-Stokes equations and viscous MHD equations. *Phys Lett A*, pages 281–286, 2003.

[6] Ghrist R Etnyre J. An index for closed orbits in Beltrami field, 2001. Available at: `http://arxiv.org/abs/math/010109`.

[7] Marsh GE. *Helicity and Electromagnetic Field Topology*. World Scientific, 1995.

[8] Freed DS. The Geometry of Loop Groups, 1985.

[9] N. Hitchin. Kählerian twistor spaces. *Proc London Math Soc*, 8(43):133–151, 1981.. Available at: `http://tinyurl.com/pb8zpqo`.

[10] Zeeman EC. *Catastrophe Theory*. Addison-Wessley Publishing Company, 1977.

[11] Nottale L Da Rocha D. Gravitational Structure Formation in Scale Relativity, 2003. Available at: http://arxiv.org/abs/astro-ph/0310036.

[12] Mills R et al. Spectroscopic and NMR identification of novel hybrid ions in fractional quantum energy states formed by an exothermic reaction of atomic hydrogen with certain catalysts, 2003. Available at: http://www.blacklightpower.com/techpapers.html.

[13] Fractional quantum Hall Effect. Available at: http://en.wikipedia.org/wiki/Fractional_quantum_Hall_effect.

[14] Fiaxat JD. A hypothesis on the rhythm of becoming. *World Futures*, 36:31–36, 1993.

[15] Fiaxat JD. The hidden rhythm of evolution, 2014. Available at: http://byebyedarwin.blogspot.fi/p/english-version_01.html.

[16] Blackman CF. *Effect of Electrical and Magnetic Fields on the Nervous System*, pages 331–355. Plenum, New York, 1994.

[17] Pitkänen M. TGD and Astrophysics. In *Physics in Many-Sheeted Space-Time: Part II*. Available at: http://tgdtheory.fi/pdfpool/astro.pdf, 2019.

[18] Pitkänen M. Massless states and particle massivation. In *p-Adic Physics*. Available at: http://tgdtheory.fi/pdfpool/mless.pdf, 2006.

[19] Pitkänen M. About the Nottale's formula for h_{gr} and the possibility that Planck length l_P and CP_2 length R are related. In *Hyper-finite Factors and Dark Matter Hierarchy: Part I*. Available at: http://tgdtheory.fi/pdfpool/vzerovariableG.pdf, 2019.

[20] Pitkänen M. Criticality and dark matter: part II. In *Hyper-finite Factors and Dark Matter Hierarchy: Part I*. Available at: http://tgdtheory.fi/pdfpool/qcritdark2.pdf, 2019.

[21] Pitkänen M. p-Adic Numbers and Generalization of Number Concept. In *TGD as a Generalized Number Theory: Part I*. Available at: http://tgdtheory.fi/pdfpool/padmat.pdf, 2019.

[22] Pitkänen M. p-Adic Physics: Physical Ideas. In *TGD as a Generalized Number Theory: Part I*. Available at: http://tgdtheory.fi/pdfpool/phblocks.pdf, 2019.

[23] Pitkänen M. Some questions related to the twistor lift of TGD. In *Towards M-Matrix: Part II*. Available at: http://tgdtheory.fi/pdfpool/twistquestions.pdf, 2019.

[24] Pitkänen M. TGD as a Generalized Number Theory: Infinite Primes. In *TGD as a Generalized Number Theory: Part I*. Available at: http://tgdtheory.fi/pdfpool/visionc.pdf, 2019.

[25] Pitkänen M. The Recent View about Twistorialization in TGD Framework. In *Towards M-Matrix: Part II*. Available at: http://tgdtheory.fi/pdfpool/smatrix.pdf, 2019.

[26] Pitkänen M. Zero Energy Ontology. In *Towards M-Matrix: part I*. Available at: http://tgdtheory.fi/pdfpool/ZEO.pdf, 2021.

[27] Pitkänen M. Hydrinos again. Available at: http://tgdtheory.fi/public_html/articles/Millsagain.pdf., 2016.

[28] Pitkänen M. Cold fusion, low energy nuclear reactions, or dark nuclear synthesis? Available at: http://tgdtheory.fi/public_html/articles/krivit.pdf., 2017.

ISSN: 2153-8301   Prespacetime Journal   www.prespacetime.com
Published by QuantumDream, Inc.

[29] Pitkänen M. Could categories, tensor networks, and Yangians provide the tools for handling the complexity of TGD? Available at: `http://tgdtheory.fi/public_html/articles/Yangianagain.pdf.`, 2017.

[30] Pitkänen M. Philosophy of Adelic Physics. In *Trends and Mathematical Methods in Interdisciplinary Mathematical Sciences*, pages 241–319. Springer.Available at: `https://link.springer.com/chapter/10.1007/978-3-319-55612-3_11`, 2017.

[31] Pitkänen M. Philosophy of Adelic Physics. Available at: `http://tgdtheory.fi/public_html/articles/adelephysics.pdf.`, 2017.

[32] Pitkänen M. TGD view about coupling constant evolution. Available at: `http://tgdtheory.fi/public_html/articles/ccevolution.pdf.`, 2018.

[33] Pitkänen M. TGD view about quasars. Available at: `http://tgdtheory.fi/public_html/articles/meco.pdf.`, 2018.

[34] Pitkänen M. Cosmic string model for the formation of galaxies and stars. Available at: `http://tgdtheory.fi/public_html/articles/galaxystars.pdf.`, 2019.

[35] Pitkänen M. Solar Metallicity Problem from TGD Perspective. Available at: `http://tgdtheory.fi/public_html/articles/darkcore.pdf.`, 2019.

[36] Pitkänen M. Some comments related to Zero Energy Ontology (ZEO). Available at: `http://tgdtheory.fi/public_html/articles/zeoquestions.pdf.`, 2019.

[37] Pitkänen M. Waterbridge experiment from TGD point of view. Available at: `http://tgdtheory.fi/public_html/articles/waterbridge.pdf.`, 2019.

[38] Pitkänen M. A critical re-examination of $M^8 - H$ duality hypothesis: part I. Available at: `http://tgdtheory.fi/public_html/articles/M8H1.pdf.`, 2020.

[39] Pitkänen M. A critical re-examination of $M^8 - H$ duality hypothesis: part II. Available at: `http://tgdtheory.fi/public_html/articles/M8H2.pdf.`, 2020.

[40] Pitkänen M. Could TGD provide new solutions to the energy problem? Available at: `http://tgdtheory.fi/public_html/articles/proposal.pdf.`, 2020.

[41] Pitkänen M. How to compose beautiful music of light in bio-harmony? `https://tgdtheory.fi/public_html/articles/bioharmony2020.pdf.`, 2020.

[42] Pitkänen M. About the role of Galois groups in TGD framework. `https://tgdtheory.fi/public_html/articles/GaloisTGD.pdf.`, 2021.

[43] Pitkänen M. Is genetic code part of fundamental physics in TGD framework? Available at: `https://tgdtheory.fi/public_html/articles/TIH.pdf.`, 2021.

[44] Pitkänen M. Is $M^8 - H$ duality consistent with Fourier analysis at the level of $M^4 \times CP_2$? `https://tgdtheory.fi/public_html/articles/M8Hperiodic.pdf.`, 2021.

[45] Pitkänen M. Questions about coupling constant evolution. `https://tgdtheory.fi/public_html/articles/ccheff.pdf.`, 2021.

[46] Pitkänen M. Some questions concerning zero energy ontology. `https://tgdtheory.fi/public_html/articles/zeonew.pdf.`, 2021.

[47] Pitkänen M. Three alternative generalizations of Nottale's hypothesis in TGD framework. `https://tgdtheory.fi/public_html/articles/MDMdistance.pdf.`, 2021.

ISSN: 2153-8301   Prespacetime Journal   www.prespacetime.com
Published by QuantumDream, Inc.

[48] Pitkänen M and Rastmanesh R. The based view about dark matter at the level of molecular biology. Available at: `http://tgdtheory.fi/public_html/articles/darkchemi.pdf.`, 2020.

Exploration

What Could 2-D Minimal Surfaces Teach about TGD?

Matti Pitkänen [1]

Abstract

In the TGD Universe space-time surfaces within causal diamonds (CDs) are fundamental objects.

1. $M^8 - H$ duality means that one can interpret the space-time surfaces in two manners: either as an algebraic surface in complexified M^8 or as minimal surfaces in $H = M^4 \times CP_2$. $M^8 - H$ duality maps these surfaces to each other.

2. Minimal surface property holds true outside the frame spanning minimal surface as 4-D soap film and since also extremal of Kähler action is in question, the surface is analog of complex surface. The frame is fixed at the boundaries of the CD and dynamically generated in its interior. At frame the isometry currents of volume term and Kähler action have infinite divergences which however cancel so that conservation laws coded by field equations are true. The frames serve as seats of non-determinism.

3. At the level of M^8 the frames correspond to singularities of the space-time surface. The quaternionic normal space is not unique at the points of a d-dimensional singularity and their union defines a surface of CP_2 of dimension $d_c = 4 - D < d$ defining in H a blow up of dimension d_c.

In this article, the inspiration provided by 2-D minimal surfaces is used to deepen the TGD view about space-time as a minimal surface and also about $M^8 - H$ duality and TGD itself.

1. The properties of 2-D minimal surfaces encourage the inclusion of the phase with a vanishing cosmological constant Λ phase. This forces the extension of the category of real polynomials determining the space-time surface at the level of M^8 to that of real analytic functions. The interpretation in the framework of consciousness theory would be as a kind of mathematical enlightenment, transcendence also in the mathematical sense.

2. $\Lambda > 0$ phases associated with real polynomials as approximations of real analytic functions would correspond to a hierarchy of inclusions of hyperfinite-factors of type II_1 realized as physical systems and giving rise to finite cognition based on finite-D extensions of rationals and corresponding extensions of p-adic number fields.

3. The construction of 2-D periodic minimal surfaces inspires a construction of minimal surfaces with a temporal periodicity. For $\Lambda > 0$ this happens by gluing copies of minimal surface and its mirror image together and for $\Lambda = 0$ by using a periodic frame.

   A more general engineering construction using different basic pieces fitting together like legos gives rise to a model of logical thinking with thoughts as legos. This also allows an improved understanding of how $M^8 - H$ duality manages to be consistent with the Uncertainty Principle (UP).

4. At the physical level, one gains a deeper understanding of the space-time correlates of particle massivation and of the TGD counterparts of twistor diagrams. Twistor lift predicts M^4 Kähler action and its Chern-Simons implying CP breaking. This part is necessary in order to have particles with non-vanishing momentum in the $\Lambda = 0$ phase.

[1] Correspondence: Matti Pitkänen http://tgdtheory.fi/. Address: Rinnekatu 2-4 8A, 03620, Karkkila, Finland. Email: matpitka6@gmail.com.

1   Introduction

In the quantum TGD based on zero energy ontology (ZEO) space-time surfaces within causal diamonds (CDs) are fundamental objects [22, 30]. $M^8 - H$ duality plays a central role: the earlier views can be found in [17, 18, 19] and the recent view in [23, 24, 27] differing in some aspects from the earlier view. $M^8 - H$ duality means that one can interpret the space-time surfaces in two manners: either as an algebraic surfaces in complexified M^8 or as minimal surfaces in $H = M^4 \times CP_2$ [30]. $M^8 - H$ duality maps these surfaces to each other.

The twistor lift of TGD is another key element [12, 13]. It replaces space-time surfaces with their 6-D twistor spaces represented as 6-D surfaces in the product of twistor spaces assignable to M^4 and CP_2 and having an induced twistor structure. This implies dimensional reduction of a 6-D Kähler action to a sum of a 4-D Kähler action and volume term having interpretation in terms of cosmological constant Λ. Kähler structure exists only for the twistor spaces of M^4 and CP_2 [1] so that the theory is unique.

Each extension of rationals (EQ) corresponds to a different value $\Lambda > 0$. For $\Lambda = 0$, the finite-D extension of rationals determined by real polynomials would be replaced with real analytic functions or subset of them.

Whether $\Lambda = 0$ can be accepted physically, will be one of the key topics of this article. At the level of adelic theory of cognition [15, 16] this question boils down to the question whether cognition is always finite and related to finite-D extensions of rationals of whether also infinite-D extensions and transcendence can be allowed.

1.1   Basic notions

$M^8 - H$ duality and twistor lift of TGD are the basic notions relevant for what follows and its is appropriate to discuss them briefly.

1.1.1   Space-time surfaces at the level of M^8

The recent view of $M^8 - H$ duality [23, 24, 27] deserves a brief summary.

At M^8 level, space-time surfaces can be regarded as algebraic 4-surfaces in complexified M^8 having interpretation as complexified octonions. The dynamical principle states that the normal space of the space-time surface at each point is associative and therefore quaternionic. The space-time surfaces are determined by the condition that the real part of an octonionic polynomial obtained as an algebraic continuation of a real polynomial with rational coefficients vanishes.

This gives a complex surface which is minimal surface from which one takes a real part by projecting to real part of complexified M^8: it is not clear whether it is minimal surface of M^8. Minimal surface property is the geometric analog of a massless d'Alembert equation [14, 20].

Also real analytic functions can be considered [23, 24] but this leads to infinite-D extensions of rationals in the adelization requiring that also the p-adic counterparts of the space-time surfaces exist. Whether this phase which corresponds to $\Lambda = 0$, where Λ is cosmological constant, can be accepted physically, will be one of the key topics of this article.

The conditions defining the space-time surfaces are exactly solvable and the conjecture is that these surfaces are minimal surfaces by their holomorphy (the induced metric of the space-time surface does not however play any role and its role is taken by the complexification number theoretic octonion norm which is real valued for the real projections) [23, 24, 27].

1.1.2   Space-time surfaces at the level of $H = M^4 \times CP_2$

At the level of $H = M^4 \times CP_2$, space-time surfaces are preferred extremals (PEs) of an action defined a variational principle fixed by the twistor lift of TGD [13]. As explained, the action consists of a volume term having interpretation in terms of length scale dependent cosmological constant Λ and Kähler action.

Physically "preferred" means holography: to a given 3-surface at the either boundary of CD one can assign a unique space-time surface as an analog of Bohr orbit. This assumption is very probably too strong: the number of Bohr orbits is finite and the dynamically determined frames of the space-time surface would characterize the non-determinism [30]. "Preferred" has several mathematical meanings, which are conjectured to be equivalent.

One of those meanings is that space-time surfaces simultaneous extremals of both volume term and Kähler action and field equations reduce almost everywhere to the analogs of the conditions satisfied by complex surfaces of complex manifolds. Note that the field equations express local conservation laws for the isometries of $H = M^4 \times CP_2$ and are in this sense hydrodynamic.

The field equations for preferred extremals do not depend on coupling parameters. This expresses quantum criticality and reduces the number of solutions dramatically as required by the fact that at the level the field equations are algebraic rather than differential equations.

Space-time surfaces are therefore minimal surfaces everywhere except at singularities, which are lower-dimensional surfaces. At singularities they are satisfied only for the entire action. The divergences of the isometry currents for the volume term and Kähler action would have delta function singularities, which must cancel each other to guarantee conservation laws.

The singular surfaces can be wormhole throats as boundaries of CP_2 type extremals at which the signature of the induced metric changes, partonic 2-surfaces acting as analogs of vertices at which light-like partonic orbits representing the lines of generalized Feynman (or twistor) diagram meet, and string world sheets having light-like boundaries at partonic orbits.

Also 3-D singularities are predicted and could be associated to time= constant hyperplanes of M^4, which in M^8 picture are associated with the roots of the polynomials determining space-time region: I have christened these roots "very special moments in the life of self" [28]. The roots define 6-spheres as universal special solutions and they intersect future light-cone along $t = r_n$ hyper-plane. It is possible to glue different solutions together along these planes so that they can serve as loci of classical non-determinism.

The singular surfaces are analogous to the frames of soap films [30]: part of them are fixed and at the boundaries of CD and part of them are dynamically generated. Classical conservation laws for the isometry currents expressing field equations pose strong conditions on what can happen in vertices.

1.1.3 $M^8 - H$ correspondence for the singularities

By $M^8 - H$ correspondence, the singular surfaces of $X^4 \subset H$ correspond to the singularities of the pre-image at the level of M^8. For the singularities $X^4 \subset M^8$ the quaternionic normal space of X^4 is not unique at points of a $d < 4$ dimensional surface but is replaced with a union of quaternionic normal spaces labelled by the points of sub-manifold of CP_2 for which the dimension is $d_c = 4 - d$. At the level of H, the singular points blow-up to d_c-dimensional surfaces. What happens for the normal space at a puncture of 3-space serves as a good analog.

In particular, the deformation of a CP_2 type extremal as a singularity corresponds to an image of a 1-D singularity with $(d = 1, d_c = 3)$ and $d_c = 3$-dimensional blow up. The properties of CP_2 type extremals suggest the 1-D curve is light-like curve for mere Kähler action and light-like geodesic for the Kähler action plus volume term.

These situations correspond to $\Lambda = 0$ and $\Lambda > 0$, where Λ is length scale dependent cosmological constant as coefficient of the volume term of action.

1.2 Key questions

The basic question to be discussed in the following is what the general ideas about 2-D minimal surfaces can teach about minimal surfaces in M^8 and H, and more generally, about quantum TGD.

1.2.1   Uncertainty Principle and $M^8 - H$ duality

The interpretation of M^8 as analog of momentum space [23, 24] meant a breakthrough in the understanding of $M^8 - H$ duality but created also a problem. How can one guarantee that $M^8 - H$ duality is consistent with Uncertainty Principle (UP)? The surfaces to which one can assign well defined momentum in M^8 should correspond to the analogs of plane waves in H and geometrically to periodic surfaces.

The fact that at the level of M^8 the surfaces are algebraic surfaces defined by polynomials with rational coefficients poses therefore a problem. Periodicity requires trigonometric functions. The introduction of real anlytic functions with rational Taylor coeffients would force the introduction of infinite-D extensions of rationals and make this possible. This is however in conflict with the idea about the finiteness of cognition forming the basic principle of adelic physics [15, 16].

1.2.2   Is the category of polymials enough?

Is it possible to have periodic minimal surfaces at the level of H or at the level of both M^8 and H without leaving the category polynomials with rational coefficients?

1. Could the non-local character of the $M^8 - H$ duality in CP_2 degrees freedom miraculously give rise to periodic functions at the level of H? Or should one perhaps modify $M^8 - H$ duality itself to achieve this [27].

2. Periodic frames assignable to light-like curves in M^8 as light-like curves would allow to achieve periodicity in the same manner as for helicoid but this requires the extension of the category of real polynomials to real analytic functions in M^8. One could even give up the assumption about a Taylor expansion with rational coefficients and assume that the coefficients belong to some possibly transcendental extension of rationals. This option would make sense in $\Lambda = 0$ phase.

3. Or could geometry come in rescue of algebra? Could one construct periodic surfaces both at the level of M^8 and H purely geometrically by gluing minimal surfaces together to form repeating patterns as is done for 2-D minimal surfaces? This option could work in $\Lambda > 0$ phases: smoothness at the junctions would be given up but local conservation laws would hold true for the entire action rather than for volume term and Kähler action separately.

1.3   Is also $\Lambda > 0$ phase physically acceptable?

Can one allow also $\Lambda = 0$ phase for the action. In this case the action reduces to mere Kähler action defined by M^4 and CP_2 Kähler forms analogous to self-dual covariantly constant $U(1)$ gauge fields? Could one see $\Lambda = 0$ phase as an analog of Higgs=0 phase?

In this phase the category of rational functions would expand to a category of real analytic functions and infinite extensions of rationals containing transcendental numbers would be unavoidable and allow light-like curves as frames instead of piecewise light-like geodesics.

One could argue that since the evolution of mathematical consciousness has led to the notion transcendentals and transcendental functions, they must be realized also at the level of space-time surfaces.

One can invent objections against the $\Lambda = 0$ phase for which Kähler action has only CP_2 part and serving at the same time as arguments for the necessity of M^4 part.

1. For a mere CP_2 Kähler action, the CP_2 type extremals representing building bricks of elementary particles become vacuum extremals and are lost from the spectrum. However, also the M^4 part of Kähler action predicted by the twistor lift gives rise to Chern-Simons (C-S) term assignable to the light-like 3-surface X_L^3 as the orbit of partonic 2-surface and one can assign a momentum to X_L^3. The boundary conditions guaranteeing momentum conservation make possible momentum exchange between interior and X_L^3.

2. CP_2 Kähler action has a huge vacuum degeneracy since space-time surfaces with 2-D Lagrangian manifold as a CP_2 projection are vacuum extremals. $\Lambda > 0$ eliminates most of these extremals. Also the M^4 part of Kähler action, which vanishes for canonically imbedded M^4, implies that most vacuum extremals of CP_2 Kähler action cease to be extremals even for $\Lambda = 0$.

3. The Kähler form of M^4 has an electric part, which is imaginary so that the energy density is of form $-E^2 + B^2$ ($= 0$ for M^4). For instance, solutions of $M^2 \times Y^2$, where Y^2 is any Lagrangian manifold of CP_2 would have negative energy for $\Lambda = 0$.

   Is the condition $\Lambda > \Lambda_{min}$ needed to exclude these extremals? Or does the PE property, requiring that CP_2 projection is a complex 2-surface, come in rescue: for a complex surface Y^2, the CP_2 magnetic energy can make the total energy positive.

2 About 2-D minimal surfaces

A brief summary about 2-D minimal surfaces and questions raised by them in TGD framework is in order. One can classify minimal surfaces to those without frame and with frame.

2.1 Some examples of 2-D minimal surfaces

The following examples about minimal surfaces are collected from the general Wikipedia article about minimal surface `https://cutt.ly/Hn673ry`) and various other Wikipedia articles. This article gives also references to articles (for instance the article "The classical theory of minimal surfaces" of Meeks and Perez [5]) and textbooks discussing minimal surfaces , see for instance [4]. Also links to online sources are given. "Touching Soap Films - An introduction to minimal surfaces" `https://cutt.ly/dmwMnJ7`) serves as a general introduction to minimal surfaces). There is also a gallery of periodic minimal surfaces (`https://cutt.ly/RmwMQ49`), which is of special interest from the TGD point of view.

2.1.1 Minimal surfaces without frame

In E^3 frameless minimal surfaces have an infinite size and are often glued from pieces, which asymptotically approach a flat plane.

Catenoid (`https://cutt.ly/in675Z6`) is obtained by a rotation of a catenoid, which is the form of the chain spanned between poles of equal height in the gravitational field of Earth. Catenoid has two planes as asymptotics and is obtained from torus by adding two punctures. Costa's minimal surface (`https://cutt.ly/in65wyP`) is obtained from torus by adding a single puncture and its second end looks like a catenoid.

Frameless minimal surfaces in E^3 allow also lattice-like structures. Schwarz minimal surface (`https://cutt.ly/dn65rJm`) is an example about minimal giving rise to 3-D lattice like structure. These surfaces have minimal genus $g = 3$.

In compact spaces closed minimal surfaces are possible and some quite surprising results hold true, see the popular article *Math Duo Maps the Infinite Terrain of Minimal Surfaces* (`http://tinyurl.com/yyetb7c7`). These surfaces have area proportional to volume of the imbedding space and the explanation is that these surfaces fill the volume densely [2, 3].

2.1.2 Minimal surfaces with lattice like structure

There exists also minimal surfaces with lattice-like structure.

1. Riemann described a one parameter of minimal surfaces with a 1-D lattice structure consisting of shelfs connected by catenoids (`https://cutt.ly/Pn65y3f`).

2. Scherk surfaces (`https://cutt.ly/3n65oeB`) are singly or doubly periodic. Scwartz surfaces (`https://cutt.ly/un65pCK`) are triply periodic structures defining 3-D lattices and have minimal genus $g = 3$. This kind of surfaces have been used to model condensed matter lattices. These surfaces have also hyperbolic counterparts.

2.1.3 Minimal surfaces spanned by frames

Minimal surfaces with frames allow to models soap films and are obtained as a solution of the Plateau's problem (`https://cutt.ly/7n65fgT`).

1. Helicoid (`https://cutt.ly/Wn65jgT`) represents a basic example of a simply periodic framed surface. Also helicoid involves transcendental functions. A portion of helicoid is locally isometric to catenoid.

2. Arbitrary curves can serve as frames with some mild restrictions. The minimal surface need not be unique. A given 2-D minimal surface is obtained in topological sense from a compact manifold by adding a puncture to represent boundaries defined by frames or the boundaries at infinity.

2.2 Some comments on 2-D minimal surfaces in relation to TGD

The study of the general properties of 2-D minimal surfaces from the TGD perspective suggest a generalization to the TGD framework and also makes possible a wider perspective about TGD itself.

2.2.1 Frameless minimal surfaces in TGD framework

Frameless minimal surfaces in E^3 have infinite sizes since they are locally saddle like. In TGD framework, the most interesting space-time surface are expected to be framed. Despite this frameless minimal surfaces are of interest.

1. In the TGD framework the minimal surfaces could extend to infinity in time-direction and remain finite in spatial directions. The asymptotically flat 2-plane could in TGD correspond to the simplest extremals of action: M^4 and "massless extremals" (MEs); surfaces $X^2 \times Y^2$ with X^2 a string world sheet and Y^2 complex manifold of CP_2; and CP_2 type extremals with 1-D light-like curve as CP_2 projection.

   Conservation laws do not allow M^4 even in principle unless the total angular momentum and color charges vanish. Various singularities could deform flat M^4 in close analogy with point and line charges.

2. In curved compact spaces also closed minimal surfaces are possible [2, 3] (`http://tinyurl.com/yyetb7c7`). One can wonder whether CP_2 as a curved space might allow a volume-filling closed 2-D or 3-D minimal surfaces besides complex surfaces and minimal Lagrangian manifolds [20]. For $\Lambda > 0$, only complex surfaces defined by polynomials in M^8 appear in PEs. It is difficult to see how this kind of exotic structure could define a physically interesting partonic 2-surface although formally one could consider a product of string world sheet and this kind of 2-surface.

2.2.2 Minimal surfaces with lattice structure

2-D minimal surfaces in E^3 allow lattice-like structures with dimensions 1, 2 and even 3. They are are interesting also in TGD framework.

1. Scwartz surface (`https://cutt.ly/un65pCK`), call it S, allows in the TGD framework a variant of form $M^1 \times S \times S^1$, where S^1 is a geodesic sphere. Same applies to all 2-D minimal surfaces allowing a lattice structure and could be in a central role in condensed matter physics according to

TGD. Also hyperbolic variants of a lattice like structure expected to relate to the tesselations of hyperbolic 3-space can be considered and could play important role at the level of magnetic bodies (MBs) as indeed suggested [25].

2. If $\Lambda = 0$ phase is physically acceptable, it would make possible light-like curves as frames and also lattice-like minimal surfaces with periodicity forced by that of the light-like curve assignable to to CP_2 type extremal as M^8 pre-image.

   Note that $\Lambda = 0$ phase relates to $\Lambda > 0$ phase by the breaking of conformal symmetry transforming light-like curves to light-like geodesics. The interpretation of $\Lambda = 0$ phase in terms of the emergence of continuous string world sheet degrees of freedom is attractive.

   Another interpretation would be based on the hierarchy of Jones inclusions of hyper-finite factors of type II_1 (HFFs). $\Lambda > 0$ phase would define the reduced configuration space ("world of calassical worlds" (WCW)) in finite measurement resolution defined by the included HFF representing measurement resolution and $\Lambda = 0$ phase as the factor without this reduction. The approximation of real analytic functions by polynomials of a given degree would define the inclusion. This sequence of approximations would be realized as genuine physical systems ,rather than only approximate descriptions of them.

3. For $\Lambda > 0$ allowing only polynomial function, periodic smooth minimal surfaces in M^8. The construction of Schwartz surface suggests how one can circumvent this difficulty.

   Schwartz surface defines a 3-D lattice obtained by gluing together analogs of unit cells. If a region of a minimal surface intersects orthogonally a plane, the gluing of this surface together with its mirror image gives rise to a larger minimal surface and one can construct an entire lattice-like system in this way. These surfaces are not smooth at the junctions.

   In the TGD framework, one would construct lattice in time direction and the gluing would occur at edges defined by 3-D $t = r_n$ planes ("very special moments in the life of self" [28]). Local conservation laws as limits of field equations are enough and derivatives can be discontinuous at $t = r_n$ planes. The expected non-uniqueness of the gluing procedure would mean a partial failure of the strict classical determinism having a crucial role in the understanding of cognition in ZEO. This is discussed in [30].

   M^8-picture suggests a very concrete geometric recipe for constructing minimal surfaces periodic in time direction and this would make it possible to realize UP for $M^8 - H$ duality.

The general vision would be that $\Lambda > 0$ phases the periodic minimal surfaces can be constructed as piecewise smooth lattice-like structures in the category of real polynomials by using the gluing procedure whereas in $\Lambda = 0$ phase they correspond to smooth surfaces in the category of real analytic functions.

2.2.3 Minimal surfaces spanned by frames

Minimal surfaces spanned by frames are of special interest from TGD point of view.

1. In the TGD framework. Minimal surfaces are spanned by fixed frames at the boundary of CD and by dynamically generated frames in the interior of CD. The dynamically generated frames break strict determinism, which means that space-time surfaces as analogs of Bohr orbits becomes non-unique [30] and holography (for its various forms see [23, 24]) forced by the General Coordinate Invariance is not completely unique.

2. CP_2 type extremal in H would correspond to 1-D singularity in M^8 analogous to a frame assigned 2-D minimal surfaces. The physical picture suggests that this curve is a light-like curve for the Kähler action ($\Lambda = 0$) and a light-like geodesic for action involving also volume term ($\Lambda > 0$). In the first case the periodicity of the light-like curve could give rise to periodic minimal surfaces as generalization of helicoid. In the second case discretized variants could replace these curves.

3. For the minimal surfaces discussed above, polynomials are not enough for their construction and the examples involve transcendental functions like trigonometric, exponential and logarithmic functions in their definition.

   The same is expected to be true also in TGD. Should one leave the category of polynomials and allow all real analytic functions with rational Taylor coefficients? Or should one assume also the $\Lambda = 0$ phase making possible real analytic functions?

   As far as cognitive representations are involved, this would mean that cognition becomes infinite since the extensions of p-adic become infinite. Could $\Lambda = 0$ phase be associated with an expansion of consciousness, kind of enlightment, and relate to mathematical consciousness?

3 Periodic minimal surfaces with periodicity in time direction

There are several motivations for the periodic minimal surfaces.

3.1 Consistency of $M^8 - H$ duality with Uncertainty Principle

Consistency of $M^8 - H$ duality with UP is one motivation.

1. M^8 is interpreted as an analog of momentum space. $M^8 - H$ correspondence must be consistent with UP. If $M^8 - H$ correspondence in M^4 degrees of freedom involves inversion of form $m^k \rightarrow \hbar_{eff} m^k/m^2$. [23, 24, 27]. This solves the problem only partially. $M^8 - H$ correspondence should realize also the idea about plane wave as space-time counterpart of point in momentum space.

   The first guess [27] would be that the $X^4 \subset CD \subset M^8$ is mapped to a union of translates of images of CD by inverse of P^k, where is the total momentum assignable to CD. What I saw as a problem, was that this gives a lattice-like many-particle state rather than a single particle state as a counterpart of a plane wave.

   If the momentum is space-like, this is indeed the case. Therefore I proposed that the image is a quantum superposition of translates rather than their union and represents an analog of plane wave. I failed to realize that this is not the case for time-like momentum since periodicity in time direction does not mean lattice as many-particle state.

   A geometric correspondence for time-like momenta is possible after all! The problem is a concrete realization of this correspondence and here the geometric construction gluing together the analogs of unit cells to form a periodic structure in time direction suggests itself.

2. Quite concretely, one could take part of $X^4 \subset CD \subset M^8$ defining particle and construct a periodic surface with a period determined by the total time-like momentum assignable to this part of X^4. X^4 has a slicing by planes $e = e_n$ [28] assignable to 6-branes with topology of S^6 defining universal special solutions of algebraic equations. Here e_n is a root of the real polynomial defining X^4.

   One could take a piece $[e_1, ..., e_k]$ of $X^4 \subset CD$ and glue it to its time reversal in M^8 to get a basic unit cell and fuse these unit cells together to obtain a periodic structure.

   The differences $e_i - e_j$, which for M^8 correspond to energy differences, are mapped by inversion to time differences $t_i - t_j$ in H. The order of magnitude for the p-adic length scale assignable to CD in question is the same as for the largest difference for the roots as conjectured on basis of the conjecture that the p-adic length scale correspond to a ramified prime of the extension dividing $|t_i - t_j|^2$ for some pair (i, j). The p-adic prime for CD need not however be a ramified prime and one can develop an argument for how it emerges [30].

3. Rather remarkably, one can glue together portions $[t_1, ..t_r]$ and the mirror image of $[t_k, t_r]$, for any k. All possible sequences of this kind are possible! This suggests an analogy to logical reasoning:

$[t_n, t_{n+1}]$ would represent a basic step $t_n \to t_{n+1}$ in the reasoning and one could combine these steps. Could this process serve as the geometric correlate for logical thought or as engineeering at the level of fundamenta interactions?

The physicalists refusing to accept non-determinism at the fundamental level fail to realize that our technology relies on a fusion of deterministic processes and is therefore not consistent with strict determinism. Also computer programs consist of deterministic pieces.

4. There is still one open question. Does the construction of the time lattices occur only at the level of H or both at the level of M^8 and H? One can argue that the realization of the analog of inverse Fourier transform forces the construction at both sides.

3.2   Bohr orbitology for particles in terms of minimal surfaces

In TGD, space-time surfaces correspond to analogs of Bohr orbits. One should also have classical space-time analogs for ordinary bound states as Bohr orbits for particles. Atoms represent the basic example. In TGD Universe, Bohr model should be much more than mere semiclassical model. Also the geodesic orbits of particles in gravitational fields should have minimal surface analogs.

The Bohr orbits should be representable as parts of minimal surfaces identifiable as deformed CP_2 type extremals. There are two options to consider corresponding to $\Lambda = 0$ phase and to $\Lambda > 0$ phases.

3.2.1 $\Lambda = 0$ phase

$\Lambda = 0$ phase corresponds to a long length scale limit but general consideratons encourage its inclusion as a genuine phase. Its relation to $\Lambda > 0$ phases would be like the relation of real numbers to extensions of rationals and transcendental functions to polynomials.

1. For $\Lambda = 0$, CP_2 type extremals are vacuum extremals and correspond to 1-D singularities, which are light-like curves in M^8 blown up to orbits of wormhole contacts in H.

   Light-like curve as an M^4 projection of Bohr orbit of this kind can give rise to "zitterbewegung" as a helical motion with average cm velocity $v < c$. The proposal for the TGD based geometric description of Higgs mechanism realizes this zitterbewegung of CP_2 type extremals for Kähler action. This makes it possible to assign to any particle orbit - be it Bohr orbit in an atom or a geodesic path in a gravitational field, an average of a light-like curve.

2. Light-likeness gives rise to Virasoro conditions emerging in the bosonic string theories. This served as a stimulus leading to the assignment of extended Kac-Moody symmetries to the light-like partonic orbits X^3 . The isometries of H define the extended Kac-Moody group. The generators of the Kac-Moody algebra depend on the complex coordinate z of the partonic 2-surface and on the light-like radial coordinate of X^3. Super-symplectic symmetries assigned to the light-like $\delta M^4_{\pm} \times CP_2$ and identified as isometries of WCW have an analogous structure [10, 26].

   The light-like orbits of the partonic 2-surfaces in H are connected by string world sheets. The interpretation could be that in $\Lambda = 0$ phase strings emerge as additional degrees of freedom.

3. For CP_2 part of Kähler action $\Lambda = 0$ CP_2 type extremals are vacua (this need not be the case for the deformations). The C-S term for CP_2 Kähler action carries no momentum and cannot contribute to momentum and cannot realize momentum conservation for deformed CP_2 type extremals.

   However, the C-S term for the M^4 part of Kähler action defines the partonic orbits as dynamical entities. If the projection of the deformation of CP_2 type extremal at the wormhole throat has M^4 projection with dimension $D = 3$, M^4 C-S term gives rise to non-vanishing momentum currents and the smooth light-orbit is consistent with the momentum conservation if boundary conditions are realized. What is remarkable that M^4 C-S term also gives rise to small CP breaking, whose origin is not understood in the standard model. The tiny C-S breaking term would be paramount for the existence of elementary particles!

The implications of this picture are rather profound. It could be possible to assign to any physical system rather detailed view about the minimal surfaces involved both at the level of H and M^8.

The basic objection against the $\Lambda = 0$ phase is that the presence of M^4 Kähler action can make energy negative.

1. The energy of surfaces of form $M^2 \times Y^2 \subset M^4 \times CP_2$, Y^2 Lagrangian 2-manifold, is negative since the electric part of M^4 Kähler form is imaginary. Can one avoid this problem?

   These Lagrangian manifolds could be required to be minimal surfaces as an outcome of a limiting procedure. They need not be PEs since they are not complex sub-manifolds: if so, the problem disappears. Indeed, if Y^2 is complex 2-surface, the CP_2 contribution to the energy can make the total energy non-negative.

2. One can also consider a minimal surface type $M^3 \times S^1 \subset M^4 \times CP_2$. This extremal could correspond in M^8 a co-dimension $d_c = 1$ singularity such that the normal planes at a given point of M^3 span geodesic circle $S^1 \subset CP_2$. Even more general curve than S^1 would be allowed by CP_2 Kähler action. This surface is a minimal surface and the energy is negative.

   Also these surfaces could be excluded by the failure of the PE property since one cannot regard M^3 as an analog of a complex surface of M^4 or S^1 as a complex surface of CP_2.

   CD in M^8 does not allow negative energies so that also $M^8 - H$ duality and quantum classical correspondence exclude negative energy extremals.

What if the negative energy PEs were possible? The energies of positive and negative energy minimal surfaces could sum up to a tachyonic momentum. What is intriguing, is that in the TGD framework p-adic mass calculations [11] require a tachyonic ground state and the massless ground states are constructed from it. Note that the tachyonic character of Higgs in the phase without symmetry breaking is analogous to this. Tachyons are also the problem of superstring theories solvable only in critical dimensions.

Could tachyonic states appear as parts of non-tachyonic states somewhat like tachyonic virtual particles appear in Feynman graphs?

1. The possibly existing periodic minimal surfaces with tachyonic total momenta would have an interpretation as lattice-like many-particle states. This excludes them as unphysical. In fact, one cannot construct tachyonic periodic minimal surfaces in the proposed way since the planes $t = t_n$ have time-like normal.

2. M^8 picture allows to interpret tachyonicity as a trick. In the M^8 picture the choice of $M^4 \subset M^8$ is in principle free. The mass squared of the particle depends on this choice since M^4 momentum is a projection of M^8 momentum to $M^4 \subset M^8$. For eigenstates of M^4 mass, one can rotate $M^4 \subset M^8$ in such a manner that the mass squared vanishes. For a superposition of states with different mass squared possible in ZEO this is not possible but one can choose M^4 so that mass squared is minimized. This gives rise to p-adic thermodynamics as a description for the mixing with heavier states.

   One could understand the tachyonic ground state as an effective description for the choice of M^4 in this manner.

3. This seems convincing to me but one cannot completely exclude the possibility of minimal surfaces with negative energy. Suppose one has a positive energy state with mass of order CP_2 mass as a state of a super-symplectic representation. Could one construct the needed tachyonic ground state using negative and positive energy space-time surfaces and combine this state with a state with mass of order CP_2 mass to get a massless state?

3.2.2 $\Lambda > 0$ phase

For $\Lambda > 0$ only light-like geodesics are possible and this forces a modification of the above picture by replacing light-like curves with piece-wise light-like geodesics.

1. A discrete variant of zitterbewegung consisting of pieces of light-like geodesics is suggestive. The dynamics in stringy degrees of freedom would be almost frozen and completely dictated by the ends of the string. Discretized version of smooth dynamics would be in question. This kind of phenomenological model for hadronic strings has been proposed.

2. The change of the direction of the partonic orbit takes place in a vertex. In M^8 picture it is associated with a partonic 2-surface associated with a $t = r_n$ hyperplane at which several CP_2 type extremals meet at the level of H. These reactions could be seen as ordinary particle reactions.

3. Another way to change the direction would be based on the interaction of parton with the interior degrees of freedom so that conservation laws are not lost. The interaction between the 3-D orbit of wormhole throat and interior is defined by the condition that normal components of the isometry currents of the total Kähler action are equal to the divergences of C-S currents the partonic orbit. For the M^4 part of C-S action only momentum currents are non-vanishing whereas for CP_2 only color currents are non-vanishing.

   At the turning points the normal current of the entire Kähler action - and the divergence of the isometry current for C-S part CP_2 type extremal must become non-vanishing and divergent but cancel each other. Local conservation laws hold true and one can speak of a momentum exchange between interior and wormhole throat. This picture applies also to color currents.

3.2.3 A connection with Higgs mechanism

The fact that zitterbewegung makes the particle effectively massive in long enough scales, suggests an analogy with the massivation by the Higgs mechanism.

1. The interactions between partonic orbits and the interior of the space-time surface are analogous to the interactions of particles with a Higgs field leading to the massivation as the Higgs field develops a vacuum expectation value.

2. M^4 Kähler form represents a constant self-dual Abelian gauge field. Although this field is not a scalar field, it is analogous to the vacuum expectation value of the Higgs field as far as its effects are considered.

3. The electric part is proportional to the imaginary unit and means that the electric contribution to the energy is negative. The way to avoid negative energies has been already considered.

3.2.4 A connection twistor diagrams and generalization of cognitive representations

Also a connection with twistor diagrams is suggestive. The light-like geodesic lines appearing as 1-D singularities in M^8 would correspond to light-like differences of the time-like momenta assignable to vertices. In H they are assignable with partonic 2-surfaces identifiable as boundaries of 3-D blow ups of 1-D singularities in M^8. In M^8, the graphs containing time-like momenta connected by singular lines would define analogs of twistor diagrams. Also at the level of H the lines connecting partonic 2-surfaces would be light-like as also the distances between them since the inversion map preserves light-likeness of the tangent curves.

   This would pose additional conditions on cognitive representations.

ISSN: 2153-8301   Prespacetime Journal   www.prespacetime.com
Published by QuantumDream, Inc.

1. The original proposal [17, 18, 19] was that cognitive representation consists of points of X^4 for which M^8 coordinates belong to the EQ associated with the polynomial considered. The expectation was that one has a generic situation so that this set is automatically finite.

   The explicit solution of the polynomial equations however led to a surprising finding was that the number of these points was a dense set for the space-time surfaces satisfying co-associativity conditions [23, 24]. The second surprise was that co-associativity (associativity of normal space) is the only possible option.

2. The additional conditions guaranteeing that the cognitive representation consists of a finite number of objects, generalize it from a discrete set of points to a union of singularities with co-dimension $d_c = 4 - d$, $d = 1, 2, 3$.

   The vertices would be connected by $d = 1$ light-like singularities and belong to 2-D partonic 2-surfaces as $d = 2$ singularities at $t = r_n$ surfaces in turn defining $d = 3$ singularities. Also 2-D string world sheets having $d = 1$ singularities as boundaries would be included.

3. This would also generalize twistor diagrams as a frame holographically coding for the space-time surface as an analog of Bohr orbit. At the M^8 level, the definition of the parts of this structure would involve only parameters with values in EQ (say the end points of a light-like geodesic defining it).

3.3   Periodic self-organization patterns, minimal surfaces, and time crystals

Periodic self-organization patterns which die and are reborn appear in biology. Even after images, which die and reincarnate, form this kind of periodic pattern. Presumably these patterns would relate to the magnetic body (MB), which carries dark matter in the TGD sense and controls the biological body (BB) consisting of ordinary matter. The periodic patterns of MB represented as minimal surface would induce corresponding biological patterns.

The notion of time crystal [7] (`https://cutt.ly/2n65xOk`) as a temporal analog of ordinary crystals in the sense that there is temporal periodicity, was proposed by Frank Wilczeck in 2012. Experimental realization was demonstrated in 2016-2017 [9] but not in the way theorized by Wilczek. Soon also a no-go theorem against the original form of the time crystal emerged [8] and motivated generalizations of the Wilzeck's proposal.

Temporal lattice-like structures defined by minimal surfaces would be obvious candidates for the space-time correlates of time crystals.

1. One must first specify what one means with time crystals. If the time crystal is a system in thermo-dynamic equilibrium, the basic thermodynamics denies periodic thermal equilibrium. A thermodynamical non-equilibrium state must be in question and for the experimentally realized time crystals periodic energy feed is necessary.

   Electrons constrained on a ring in an external magnetic field with fractional flux posed to an energy feed form a time crystal in the sense that due to the repulsive Coulomb interaction electrons form a crystal-like structure which rotates. This example serves as an illustration of what time crystal is.

2. Breaking of a discrete time translation symmetry of the energy feed takes place and the period of the time crystal is a multiple of the period of the energy feed. The periodic energy feed guarantees that the system never reaches thermal equilibrium. According to the Wikipedia article, there is no energy associated with the oscillation of the system. In rotating coordinates the state becomes time-independent as is clear from the example. What comes to mind is a dynamical generation of Galilean invariance applied to an angle variable instead of linear spatial coordinate.

3. Also the existence of isolated time crystals has been proposed assuming unusual long range interactions but have not been realized in laboratory.

Time crystals are highly interesting from the TGD perspective.

1. The periodic minimal surfaces constructed by gluing together unit cells would be time crystals in geometric sense (no thermodynamics) and would provide geometric correlates for plane waves as momentum eigenstates and for periodic self-organization patterns induced by the periodic minimal surfaces realized at the level of the magnetic body. It is difficult to avoid the idea that geometric analogs of time crystals are in question.

2. The hierarchy of effective Planck constants $h_{eff} = nh_0$ is realized at the level of MB. To preserve the values of h_{eff} energy feed is needed since h_{eff} tends to be reduced spontaneously. Therefore energy feed would be necessary for this kind of time crystals. In living systems, the energy feed has an interpretation as a metabolic energy feed.

   The breaking of the discrete time translation symmetry could mean that the period at MB becomes a multiple of the period of the energy feed. The periodic minimal surfaces related to ordinary matter and dark matter interact and this requires con-measurability of the periods to achieve resonance.

3. Zero energy ontology (ZEO) predicts that ordinary ("big") state function reduction (BSFR) involves time reversal [22, 30]. The experiments of Minev et al [6, 29] give impressive experimental support for the notion in atomic scales, and that SFR looks completely classical deterministic smooth time evolution for the observer with opposite arrow of time. Macroscopic quantum jump can occur in all scales but ZEO together with h_{eff} hierarchy takes care that the world looks classical! The endless debate about the scale in which quantum world becomes classical would be solely due to complete misunderstanding of the notion of time.

4. Time reversed dissipation looks like self-organization from the point of view of the external observer. A sub-system with non-standard arrow of time apparently extracts energy from the environment [21]. Could this mechanism make possible systems in which periodic oscillations take place almost without external energy feed?

Could periodic minimal surfaces provide a model for this kind of system?

1. Suppose that one has a basic unit consisting of the piece $[t_1, .., t_k]$ and its time reversal glued together. One can form a sequence of these units.

   Could the members of these pairs be in states, which are time reversals of each other? The first unit would be in a self-organizing phase and the second unit in a dissipative phase. During the self-organizing period the system would extract part of the dissipated energy from the environment. This kind of state would be "breathing" [31].

   There is certainly a loss of energy from the system so that a metabolic energy feed is required but it could be small. Could living systems be systems of this kind?

2. One can consider also more general non-periodic minimal surfaces constructed from basic building bricks fitting together like legos or pieces of a puzzle. These minimal surfaces could serve as models for thinking and language and behaviors consisting of fixed temporal patterns.

Received June 28, 2021; Accepted December 4, 2021

References

[1] N. Hitchin. Kählerian twistor spaces. *Proc London Math Soc*, 8(43):133–151, 1981.. Available at: http://tinyurl.com/pb8zpqo.

[2] Marques FC Irie K and Neves A. Density of minimal hypersurfaces for generic metrics, 2018. Available at: `https://arxiv.org/pdf/1710.10752.pdf`.

[3] Marques FC Liokumovich Ye and Neves A. Weyl law for the volume spectrum, 2018. Available at: `https://arxiv.org/pdf/1607.08721.pdf`.

[4] Osserman R. *A survey of minimal surfaces. Second edition.* Dover Publications Inc., New York, 1986.

[5] Meeks WH and Perez J. The classical theory of minimal surfaces. *Bull. Amer. Math. Soc.*, 48(3), 2011. Available at: `https://cutt.ly/xmw1vUQ`.

[6] Minev ZK et al. To catch and reverse a quantum jump mid-flight, 2019. Available at: `https://arxiv.org/abs/1803.00545`.

[7] Wilczek F. Quantum Time Crystals. *Phys. Rev. Lett.*, 109(16), 2012. Available at: `https://journals.aps.org/prl/abstract/10.1103/PhysRevLett.109.160401`.

[8] Oshikawa M Watanabe H. Absence of Quantum Time Crystals. *Phys. Rev. Lett.*, 114(25), 2015. Available at: `https://journals.aps.org/prl/abstract/10.1103/PhysRevLett.114.251603`.

[9] Zhang et al. Observation of a Discrete Time Crystal, 2016. Available at:`https://arxiv.org/abs/1609.08684`.

[10] Pitkänen M. Recent View about Kähler Geometry and Spin Structure of WCW . In *Quantum Physics as Infinite-Dimensional Geometry*. Available at: `http://tgdtheory.fi/pdfpool/wcwnew.pdf`, 2014.

[11] Pitkänen M. Massless states and particle massivation. In *p-Adic Physics*. Available at: `http://tgdtheory.fi/pdfpool/mless.pdf`, 2006.

[12] Pitkänen M. Some questions related to the twistor lift of TGD. In *Towards M-Matrix: Part II*. Available at: `http://tgdtheory.fi/pdfpool/twistquestions.pdf`, 2019.

[13] Pitkänen M. The Recent View about Twistorialization in TGD Framework. In *Towards M-Matrix: Part II*. Available at: `http://tgdtheory.fi/pdfpool/smatrix.pdf`, 2019.

[14] Pitkänen M. About minimal surface extremals of Kähler action. Available at: `http://tgdtheory.fi/public_html/articles/minimalkahler.pdf.`, 2016.

[15] Pitkänen M. Philosophy of Adelic Physics. In *Trends and Mathematical Methods in Interdisciplinary Mathematical Sciences*, pages 241–319. Springer.Available at: `https://link.springer.com/chapter/10.1007/978-3-319-55612-3_11`, 2017.

[16] Pitkänen M. p-Adicization and adelic physics. Available at: `http://tgdtheory.fi/public_html/articles/adelicphysics.pdf.`, 2017.

[17] Pitkänen M. Does $M^8 - H$ duality reduce classical TGD to octonionic algebraic geometry?: part I. Available at: `http://tgdtheory.fi/public_html/articles/ratpoints1.pdf.`, 2017.

[18] Pitkänen M. Does $M^8 - H$ duality reduce classical TGD to octonionic algebraic geometry?: part II. Available at: `http://tgdtheory.fi/public_html/articles/ratpoints2.pdf.`, 2017.

[19] Pitkänen M. Does $M^8 - H$ duality reduce classical TGD to octonionic algebraic geometry?: part III. Available at: `http://tgdtheory.fi/public_html/articles/ratpoints3.pdf.`, 2017.

[20] Pitkänen M. Minimal surfaces: comparison of the perspectives of mathematician and physicist. Available at: `http://tgdtheory.fi/public_html/articles/minimalsurfaces.pdf.`, 2019.

[21] Pitkänen M. Quantum self-organization by h_{eff} changing phase transitions. Available at: `http://tgdtheory.fi/public_html/articles/heffselforg.pdf.`, 2019.

[22] Pitkänen M. Some comments related to Zero Energy Ontology (ZEO). Available at: `http://tgdtheory.fi/public_html/articles/zeoquestions.pdf.`, 2019.

[23] Pitkänen M. A critical re-examination of $M^8 - H$ duality hypothesis: part I. Available at: `http://tgdtheory.fi/public_html/articles/M8H1.pdf.`, 2020.

[24] Pitkänen M. A critical re-examination of $M^8 - H$ duality hypothesis: part II. Available at: `http://tgdtheory.fi/public_html/articles/M8H2.pdf.`, 2020.

[25] Pitkänen M. Is genetic code part of fundamental physics in TGD framework? Available at: `https://tgdtheory.fi/public_html/articles/TIH.pdf.`, 2021.

[26] Pitkänen M. Summary of Topological Geometrodynamics. `https://tgdtheory.fi/public_html/articles/tgdarticle.pdf.`, 2020.

[27] Pitkänen M. Is $M^8 - H$ duality consistent with Fourier analysis at the level of $M^4 \times CP_2$? `https://tgdtheory.fi/public_html/articles/M8Hperiodic.pdf.`, 2021.

[28] Pitkänen M. $M^8 - H$ duality and consciousness. Available at: `http://tgdtheory.fi/public_html/articles/M8Hconsc.pdf.`, 2019.

[29] Pitkänen M. Copenhagen interpretation dead: long live ZEO based quantum measurement theory! Available at: `http://tgdtheory.fi/public_html/articles/Bohrdead.pdf.`, 2019.

[30] Pitkänen M. Some questions concerning zero energy ontology. `https://tgdtheory.fi/public_html/articles/zeonew.pdf.`, 2021.

[31] Pitkänen M and Rastmanesh R. Homeostasis as self-organized quantum criticality. Available at: `http://tgdtheory.fi/public_html/articles/SP.pdf.`, 2020.

ISSN: 2153-8301   Prespacetime Journal   www.prespacetime.com
Published by QuantumDream, Inc.

Article

TGD View of the Engine Powering Jets from Active Galactic Nuclei

Matti Pitkänen [1]

Abstract

The identification of the energy source (central engine) explaining the energy loss associated with the jets from active galactic nuclei (AGNs) is a long-standing problem of astrophysics. In the model of Blandford and Znajek (BZ model) for the central engine as a blackhole, the Penrose process would provide the energy. The energy would come basically from the blackhole mass. Empirical support for the BZ model emerges from the study of the supermassive blackhole associated with a galaxy known as Messier 87 (M87). The finding is that the magnetic field associated with the jet structure is tightly wound helical structure and so strong that it would control the dynamics of the matter from falling to blackhole except by occasional leakages. Electron-positron pairs created in the annihilation of photons would accelerate in the force-free helical electromagnetic field having also an electric component. The TGD based model involves several aspects of the new physics predicted by TDG. TGD leads to a model of galaxies and other astrophysical structures. Inflaton decay is replaced with the thickening of cosmic strings to flux tubes liberating as ordinary matter. Hierarchy of Planck constants $h_{eff} = nh_0$, in particular Nottale's hypothesis predicts quantum coherence in the exterior of in scales at least of order Schwartschild radius of the blackhole-like entity. Zero energy ontology (ZEO) predicts that the arrow of time changes in ordinary state function reductions. TGD replaces black-holes with blackhole-like entities (BHs) and white-holes with their time reversals (WHs) allowed in ZEO. BH (WH) would be a volume filling flux tube but with a relatively small value of h_{eff}. In the case of WH, it would provide "metabolic energy" for jets and take care that the value of h_{eff} is preserved (the analogy with living systems is very strong). The jets would be analogous to laser beams/supracurrents with a huge value of $h_{eff} = h_{gr}$. The model would also explain the ultrahigh energy cosmic rays. The force-free fields would be generalized Beltrami fields associated with flux tubes and identifiable as minimal surfaces in the Minkowskian regions of space-time surface. The absence of classical dissipation would be a correlate for the absence of dissipation for supra-currents and dark photon laser beams.

1 Introduction

This work was inspired by a Quanta Magazine article "Physicists Identify the Engine Powering blackhole Energy Beams" (`https://cutt.ly/dQtDK7Q`) telling about an empirical support for the model of Blandford and Znajek (BZ model in the sequel) [10] for the central engine providing energy for jets from active galactic nuclei (AGNs). In the BZ model (`https://cutt.ly/9QtD0kK`) AGN is identified as a blackhole and the Penrose process would provide the energy of the jets emerging from the blackhole. The energy would come basically from the blackhole mass. The empirical support is found by studying the supermassive blackhole associated with a galaxy known as Messier 87 (M87).

The basic problem is the identification of the central engine of the active galactic nuclei (AGNs) (`https://cutt.ly/eQtFyib`) providing the huge energy feed to the the jets.

1.1 Typical properties of active galactic nuclei

The power emitted by active galactic nuclei (AGNs) is typically of the order of 10^{38} W corresponding to a transformation of a mass of 10^{22} kg per second to energy. The typical radius of the AGN is $R \simeq 2$ AU for the active region.

---

[1]Correspondence: Matti Pitkänen `http://tgdtheory.fi/`. Address: Rinnekatu 2-4 8A, 03620, Karkkila, Finland. Email: matpitka6@gmail.com.

One must distinguish between magnetic fields associated with the interior of the central objects, the region near its surface, and the jet region with the scale of visible jets about 10^5 ly. According to the estimate of [9] (https://cutt.ly/DQtFzD3), the magnetic field is about 10 Gauss in the jet region. About $10^6 - 10^7$ Gauss near horizon. Also near-horizon magnetic fields in the range $10^8 - 10^{11}$ Tesla have been proposed for some AGNs.

Quasars are examples of AGNs and also M87 central region idenfied as a blackhole is such. In this case the mass is $6.5 \times 10^9 M_{Sun}$ and Scwartschild radius is about 2×10^{10} km $= 1.3 \times 10^3$ AU. For M87 central object the magnetic field in the jet region is that of refrigerator magnet and about 100 Gauss. For Sagittarius A has a in the center of the Milky Way the radius of the central object .4 AU.

One can pose some general conditions on the central engine serving as the energy source for the jets. The time scale Δt for the luminosity fluctuations in the power should satisfy $\Delta t < R$. For M87 one has $\Delta t \leq 10^4$ s. The gravitational force is assumed to be balanced by the radiation pressure of the outward radiation.

Consider now the observations about the M87 blackhole-like entity (https://cutt.ly/dQtDK7Q).

1. The mass of the M87 black-hole-like entity is about $6.5 \times 10^9 M_{Sun}$.

2. There are two 5,000 ly long white hot plasma jets travelling in opposite time directions with emitted power of 3×10^{36} W. They have blobs at their ends. Synchrotron radiation is emitted at radio wavelengths in the magnetic field, which according to the popular article has the strength of a refrigerator magnet, so that one would have $B \sim 100$ Gauss. Both the intensity of B and the size of the emitting region contribute to the intensity of the energy flow.

3. There are two alternatives for the BZ process that have been developed and explored in hundreds of computer simulations in recent decades. They have acronyms MAD and SANE.

   For the SANE option B is weak: charged matter dominates over B. For the MAD option B is strong, has a spiral structure and acts as a "boss" of the matter. The tight spiral structure forms a sleeve around the jet preventing charges from entering the central object. This inspires a critical question: doesn't the object look more like a whitehole.

   The strongly polarized light in the Event Horizon Telescope's new photo suggests strong magnetic fields, and supports the MAD version. B has a strength of about 100 Gauss, that is 200 times the strength of the Earth's magnetic field with the nominal value $B_E \simeq .5$ Gauss. The polarization pattern for the radio waves is found to be stripy and the polarization in a plane locally: this allows us to conclude that the magnetic field is indeed helical and non-random.

1.2   Central engine as a Penrose process?

In the BZ model, the central object is assumed to be a blackhole and Penrose process would provide the energy feed to the jets of length about 5000 ly. Note that the Milky Way is about 1,0000 ly thick.

1. The blackhole is surrounded by an accretion disk from which the matter ends down to the BH.

2. Kerr solution of Einstein-Maxwell field equations [3, 7] (https://cutt.ly/sQuKXth) involving magnetic field is the starting point. Matter falling into the Kerr blackhole rotates and the magnetic field lines are twisted to helical shape. By Faraday law, an electric field along field lines is generated by the rotation of the flux lines. Electrons and positrons created in the annihilation photons emitted as the particles fall to the region near the blachole, start to flow along the field lines of the electric field in opposite directions and generate the jets.

3. The model assumes that the electromagnetic field is force free so that it does not dissipate and Lorentz force vanishes. At a single particle level this implies the condition $E + qv \times B = 0$. Vanishing dissipation requires $v \cdot E = 0$. This helical structure would be in the direction of the jet.

4. The basic question has been whether it is accretion disk or magnetic field that controls the dynamics. The first option, known as SANE, corresponds to weak and incoherent magnetic fields. The second option, known as MAD, corresponds to strong and coherent magnetic fields.

5. MAD is favoured by the recent observations. Magnetic field would form a sleeve around the jet and the synchrotron radiation pressure would prevent matter from falling into the blackhole. Matter can only occasionally leak to blackhole.

One can however wonder whether it makes sense to talk about blackhole anymore! Doesn't this look more like white hole as a time reversal of blackhole feeding energy and matter to the environment?

1.3   TGD inspired view of the central engine

In the TGD framework the model of the central engine as a Penrose process is replaced by the following picture. The key concepts are following:

1. Space-time is identified as a 4-D minimal surface in $H = M^4 \times CP_2$ [44] or as an algebraic surface in complexified M^8 having octonionic interpretation [33, 34, 39]. These descriptions are related by $M^8 - H$ duality analogous to momentum-position duality, which does not generalize from wave mechanics to quantum field theory (QFT). Therefore the points or M^8 are 8-momenta.

   The classical dissipation is absent for the generalized Beltrami fields and the proposal [44] is that minimal surfaces (apart form singularities defining dynamically generated frame for space-time surfaces as analog of a soap film) define locally generalized Beltrami fields.

2. Zero energy ontology (ZEO) [32, 42, 40] predicts that time reversal occurs in the TGD counterparts of ordinary state function reductions ("big" SFRs) but not in "small" SFRs (SSFRs).

3. The hierarchy of effective Planck constants predicts a hierarchy of phases of ordinary matter labelled by the values of effective Planck constant $h_{eff} = nh_0$. The phases with different values of h_{eff} behave in many respects like dark matter with respect to each other. The findings of Randell Mills suggest $\hbar/\hbar_0 = 6$ [23] but also larger values for this ratio can be considered [41]. In [41] it is proposed that the $\hbar_0/\hbar$ is equal to the ratio l_P/R of Planck length l_P to CP_2 radius R.

   As a special case, one obtains gravitational Planck constant satisfying $h_{eff} = h_{gr} = GMm/\beta_0$, where $\beta_0 = v_0/c$ and $\beta_0 < c$ has dimensions of velocity, as a generalization of Nottale's hypothesis [8]. The gravitational Compton length $\lambda_{gr} = \hbar_{gr}/m = GM/\beta_0$ does not depend on m and is equal to Schwartschild radius r_s for $\beta_0 = 1/2$. Also the cyclotron energy spectrum $E_c = nGMqB/\beta_0$ is independent of the mass of the charged particle.

   The hierarchy of Planck constants, the notion of $\hbar_{gr}$, and coupling constant evolution are discussed in detail in [41].

Consider now the key elements of the model.

1. TGD leads to a general model for the formation of galaxies, stars, planets,... in terms of cosmic strings thickening to flux tubes [29, 30, 31]. The energy of the flux tube, which consists of a volume energy and Kähler magnetic energy, is transformed to ordinary matter as the string tension is reduced in a sequence of phase transitions reducing the length scale dependent cosmological constant Λ.

   This process is analogous to the decay of an inflaton field to matter. The model (there are actually several basic variants of it) explains the flat velocity spectra associated with the spiral galaxies. For the first option, a long cosmic string normal to the galactic plane causes the gravitational field explaining the flat velocity spectrum of spiral galaxies. For galaxies formed around closed flux loops the velocity spectrum is not flat. There is no dark matter halo although it is possible that the galactic plane contains cosmic strings parallel to the plane.

2. ZEO (ZEO) [32, 42, 40], which predicts that the TGD counterparts of ordinary state function reductions (SFRs) involve time reversal, is involved in an essential manner. TGD predicts both blackhole-like objects (BH) and whitehole-like objects (WH) as the time reversals of BHs. The seed of the galaxy, active galactic nucleus (AGN), involves WH. Quasars are cases of AGNs as WHs [29, 30].

3. In the TGD framework, the Kerr blackhole [3, 7] is replaced with a whitehole-like object (WH). Kerr blackhole indeed has an opposite arrow of time reversal as the distant environment. The WH is time reversal of BH and feeds matter and energy to the environment. This serves as an analog of the Penrose process in the TGD based model.

4. The TGD analog for the rotation of spacetime and the twisting of the magnetic field lines near the Kerr blackhole is very concrete. Space-time is a 4-surface and the flux tubes carrying monopole flux are pieces of 3-space as a 3-surface. They quite concretely rotate and get twisted in the process. Analogous process occurs in the Sun with a period of 11 years ending as reconnections untwist the flux tubes.

5. WH would correspond to a tangle of a long cosmic string in the direction of the jet thickened to a flux tube but still carrying an extremely strong magnetic field. The helical magnetic field in the exterior of the jet would not represent return flux of this field as one might first think. There is a current ring associated with the equator of Earth, which carries a parallel magnetic field analogous to the helical magnetic field.

   The magnetic field in the exterior of WH is associated with a space-time surface, which is many-sheeted with respect to CP_2 rather than M^4 so that either CP_2 or cosmic string world sheet $M^2 \times S^2 \subset M^4 \times CP_2$) would serve as the arena of physics rather than M^4, which is quantum coherent flux tube bundle analogous to BE-condensate. M^4 coordinates as functions of CP_2 or $M^2 \times CP_2$ coordinates would be many-valued rather than vice versa. This picture is very natural if one accepts $M^8 - H$ duality [33, 34, 39].

   Cosmic strings dominate during the primordial cosmology in TGD Universe, and the analog of the inflationary period corresponds to the transition to a phase in which the Einsteinian space-time with M^4 as the arena of physics is a good approximation. Hence the $M^2 \times CP_2$ option looks more plausible.

6. The force-free em fields appearing in the BZ model [10] correspond to space-time surfaces as minimal surfaces realizing a 4-D generalization [14, 15, 13] of 3-D Beltrami fields [2, 6, 4, 5], which do not not dissipate classically [16]. The interpretation of the non-dissipating Kähler currents is as classical correlates for supracurrents [36]. The prediction is that charged particles flow without dissipation that is as supra currents: not only Cooper pairs but also charged fermions. Also the analogs of laser beams of dark photons are expected.

7. The hierarchy of Planck constants [20] [41] is an important piece of the picture emerging from adelic physics [26, 27]. From $\hbar_{gr} = GMm/\beta_0$ realizing Equivalence Principle, the gravitational Compton length $\lambda_{gr} = r_s/2\beta_0$ is universal and equals to r_S for $\beta_0 \equiv \beta_0 = 1/2$.

   All astrophysical objects are predicted to be quantum coherent in the scale of $\lambda_{gr} = r_s/2\beta_0$ at least. The quantum coherence would be at the level of magnetic body (MB) [31, 30]. WH/BH as a thickened flux tube tangle would not have large h_{eff} but would be accompanied by a large scale quantum object.

   The astroscopic quantum coherence would be associated with the helical magnetic field surrounding the long cosmic string having BH or WH as a tangle.

8. Also the cyclotron energy spectrum is universal and does not depend on the mass of the charged particles so that all charged particles rather than only electrons are expected to form supracurrents.

Dark matter would flow along flux tubes and form the dark core of the jet, perhaps extending over cosmic distances to other galaxies identified as tangles of the one and the same cosmic string.

Stars and even planets would be parts of this fractal network. Dark cyclotron states have huge energies for $h_{eff} = h_{gr}$ serving also as a measure for algebraic complexity and, in the TGD inspired theory of consciousness, also for intelligence and scale of quantum coherence. The analogy with a cosmic nervous system is obvious.

9. The decay of quantum coherent states to ordinary states takes place by the loss of quantum coherence in which $h_{eff} = h_{gr}$ is reduced. This would create the visible jets and blobs at their ends. For M87, which is elliptical for which the velocity spectrum is not flat, the flux tubes would be closed in a relatively short scale. Their length scale could be that of the jets in the case of ellipticals. The thickening of the cosmic string at the core leads to the reduction of mass of WH and gives rise to the flow of mass and energy to the environment. One could see this process as a time reversal for the generation of BH and perhaps also as an analogy for the evaporation of BH.

10. $M^8 - H$ duality [33, 34, 39] and adelic physics [27] help to understand the decoherence process geometrically. The reduction of h_{eff} and thus of the length scale of quantum coherence, allows a number theoretic description at the level of M^8. An irreducible polynomial, which depends on parameters, reduces to a product of polynomials for some critical values of the parameters. This gives rise to a set of disjoint space-time surfaces, which are not correlated. This means decoherence. This includes as a special case the description of catastrophic changes in catastrophe theory of Thom [1]. The maximal decoherence produces a product of first order polynomials with rational roots.

At the level of H this corresponds to a decay of coherent flux tube bundle to disjoint uncorrelated flux tubes.

2 TGD based model for the formation of galaxies and other astrophysical structures

TGD leads to a rather detailed picture about the formation of galaxies and other basic astrophysical structures [29, 30, 31].

2.1 Brief description of the model for for the formation of galaxies and stars

TGD based cosmology predicts that the primordial cosmology was dominated by cosmic strings identified as 4-surfaces having 2-D M^4 projection in $H = M^4 \times CP_2$. CP_2 projection is a complex surface of CP_2. The dimension of M^4 projection is unstable against perturbations and during cosmological evolution the M^4 projection thickens. This leads to a model for the formation of galaxies as tangles along cosmic strings in turn containing stars and even planets as sub-tangles [30, 29, 31].

1. Twistor lift of TGD [17] predicts that cosmological constant Λ at the level of space-time surface (to be distinguished from that associated with GRT limit of TGD) is length scale dependent. This solves the basic problem caused by the huge value of cosmological constant in the very early Universe. In ZEO length scale dependent Λ having spectrum coming as some negative powers of 2 characterizes the space-time sheets assignable to individual system and the corresponding causal diamond (CD) and is determined by its p-adic length scale.

   For instance, Sun has its own cosmological constant predicted by the model solving the puzzle due to larger abundances obtained in solar-seismological determinations than in spectroscopic and meteoritic determinations. Dark nuclear states of nuclei inside solar core contribute also to the nuclear abundances [31].

2. The energy of flux tubes consists of Kähler magnetic energy and volume energy. Quantum classical correspondence strongly suggests that this energy is identifiable as dark matter even for minimal value of h_{eff}.

3. Phase transitions reducing the value of cosmological constant are possible. Cosmic strings (or rather their M^4 projections) start to thicken and lose magnetic energy by transforming to ordinary matter. This is analogous to the decay of the inflaton field to matter. This generates Einsteinin space-time with space-time surfaces having large and increasing 4-D M^4 projection. Flux tubes and cosmic strings are however still present.

   The expansion of flux tubes in phase transitions reducing Λ gives rise to a jerk-wise accelerated expansion at the level of astrophysical objects. For given phase transition the accelerated expansion eventually stops since the expansion increases volume energy. The expansion periods however repeat being induced by phase transitions reducing length scale dependent quantized cosmological constant Λ associated with the volume action coming as powers of 2 and making flux tubes unstable against thickening and transformation of magnetic energy to ordinary matter.

   The recent accelerated expansion corresponds to this kind of period being thus analogous to inflation and is predicted to stop since volume energy increases. The expansion rate is predicted to oscillate so that the expansion takes place as jerks and there is evidence for this [11] (see (`http://tinyurl.com/oqcn2hp`) discussed from TGD point of view in [18].

4. In particular, the TGD counterpart of inflation would have led from cosmic string dominated primordial cosmology in which Einsteinian space-time does not make sense to a radiation dominated phase in which Einsteinian space-time makes sense. Expanding Earth model [28, 43], which allows to understand Cambrian Explosion is one application of TGD based quantum cosmology.

2.2   The notion of length scale dependent cosmological constant

TGD predicts that cosmological constant Λ characterizing space-time sheets is length scale dependent and depends on p-adic length scale. Furthermore, expansion would be fractal and occur in jerks. This is the picture that twistor lift of TGD leads to [17].

Quite generally, cosmological constant defines itself a length scale $R = 1/\Lambda^{1/2}$. $r = (8\pi)^{1/4}\sqrt{Rl_P}$ - essentially the geometric mean of cosmological and Planck length - defines second much shorter length scale r. The density of dark energy assignable to flux tubes in TGD framework is given as $\rho = 1/r^4$.

In TGD framework these scales corresponds two p-adic length scales coming as half octaves. This predicts a discrete spectrum for the length scale dependent cosmological constant Λ [17, 19, 21] . For instance, one can assign to ..., galaxies, stars, planets, etc... a value of cosmological constant. This makes sense in many-sheeted space-time but not in standard cosmology.

Cosmic expansion is replaced with a sequence of fast jerks reducing the value of cosmological constant by some power of 2 so that the size of the system increases correspondingly. The jerk involves a phase transition reducing Λ by some negative power of 2 inducing an accelerating period during which flux tube thickness increases and magnetic energy transforms to ordinary matter. Thickening however increases volume energy so that the expansion eventually halts. Also the opposite process could occur and could correspond to a "big" state function reduction (BSFR) in which the arrow of time changes.

An interesting question is whether the formation of neutron stars and super-novas could involve BSFR so that these collapse phenomena would be kind of local Big Bangs but in opposite time direction. One can also ask whether blackhole evaporation could have as TGD analog BSFR meaning return to original time direction by a local Big Bang. TGD analogs of blackholes are discussed in [29].

Evidence for the anisotropy of the acceleration of cosmic expansion has been reported (see `http://tinyurl.com/rx4224f`). Anisotropy of cosmic acceleration would fit with the hierarchy of scale dependent cosmological constants predicting a fractal hierarchy of cosmologies within cosmologies down to particle physics length scales and even below. The phase transitions reducing the value of Λ for given causal diamond would induce accelerated inflation like period as the magnetic energy of flux tubes decays to ordinary particles. This would give a fractal hierachy of accelerations in various scales.

2.3   Some examples

Consider now some representative examples to see whether this picture can be connected to empirical reality [30].

1. Cosmological constant in the length scale of recent cosmology corresponds to $R \sim 10^{26}$ m (see `http://tinyurl.com/k4bwlzu`). The corresponding shorter scale $r = (8\pi)^{1/4}\sqrt{Rl_P}$ is identified essentially as the geometric mean of R and Planck length l_P and equals to $r \sim 4 \times 10^{-4}$ m: the size scale of large neuron. This is very probably not an accident: this scale would correspond to the thickness of monopole flux tubes.

2. If the large scale R is solar radius about 7×10^8 m, the short scale $r \simeq 10^{12}$ m is about electron Compton length, which corresponds to p-adic length scale $L(127)$ assignable to Mersenne prime $M_{127} = 2^{127} - 1$. This is also the size of dark proton explaining dark fusion deduced from Holmlid's findings [24, 25]: this requires $h_{eff} \sim 2^{12}$!

   Remark: Dark proton sequences could be neutralized by a sequence of ordinary electrons locally. This could give rise to analogs of atoms with electrons being very densely packed along the flux tube.

   The prediction of the TGD based model explaining the 10 year old puzzle related to the fact that nuclear abundances in solar interior are larger than outside [31] (`http://tinyurl.com/y38m54ud`) assumes that nuclear reactions in Sun occur through intermediate states which are dark nuclei. Hot fusion in the Sun would thus involve the same mechanism as "cold fusion" [25, 35]. The view about cosmological constant and TGD view about nuclear fusion lead to the same prediction.

3. If the short scale is p-adic length $L(113)$ assignable to Gaussian Mersenne $M_{G,113} = (1 + i)^{113} - 1$ defining nuclear size scale of $r \sim 10^{-14}$ m, one has $R \sim 10$ km, the radius of a typical neutron star (see `http://tinyurl.com/y5ukv2wt`) having a typical mass of 1.4 solar masses.

   A possible interpretation is as a minimum length of a flux tube containing sequence of nucleons or nuclei and giving rise to a tangle. Neutron would take volume of about nuclear size - size of the magnetic body of neutron? Could supernova explosions be regarded as phase transitions scaling the stellar Λ by a power of 2 by making it larger and reducing dramatically the radius of the star?

4. Short scale $r \sim 10^{-15}$ m corresponding to proton Compton length gives R about 100 m. Could this scale correspond to quark star (see `http://tinyurl.com/y3n78tjs`)? The known candidates for quark stars are smaller than neutron stars but have considerably larger radius measured in few kilometers. Weak length scale would give large radius of about 1 cm. The thickness of flux tube would be electroweak length scale.

Starting from this picture, one ends up to rather detailed picture making correct predictions about minimum radii of blackholes and neutron stars.

1. The idea about ordinary stars as blackhole-like objects in generalized sense emerges naturally since flux tubes are universal objects in TGD Universe and could be also inspired by the fashion of dualizing everything to blackholes.

2. The standard blackhole thermodynamics is replaced by two thermodynamics. The first thermodynamics is assignable to the flux tubes as string like entities having Hagedorn temperature T_H as maximal temperature.

The second thermodynamics is assignable to the gravitational flux tubes characterized by the gravitational Planck constant h_{gr}: Hawking temperature T_B is scaled up by the ratio $\hbar_{gr}/\hbar$ to $T_{B,D}$ and is gigantic as compared to the ordinary Hawking temperature but the intensity of dark Hawking radiation is extremely low.

The condition $T_H = T_{B,D}$ for thermodynamical equilibrium fixes the velocity parameter $\beta_0 = v_0/c$ appearing in the Nottale formula for $\hbar_{gr}$ and suggests $\beta_0 = 1/h_{eff}$ for the dark nuclei at flux tubes defining star as blackhole like entity in TGD sense.

This also predicts the Hagedorn temperature of the counterpart of blackhole in GRT sense to to be hadronic Hagedorn temperature assignable to the flux tube containing dark nuclei as dark nucleon sequences so that there is a remarkable internal consistency. In ZEO (ZEO) quasars and active galactic nuclei can be seen as white-hole like objects (WHs) and time reversals of blackhole-like objects (BHs).

3. The cosmological time anomalies such as stars older than the Universe can be understood. In ZEO the time evolution for the zero energy states associated with causal diamonds (CDs) by sequences of small state function reductions (weak measurements) gives rise to conscious entity, self. Self dies and re-incarnates with an opposite arrow of time in big (ordinary) state function reduction reversing the arrow of time. These reincarnations define kind of universal Karma's cycle. If the Karma's cycle leaves the sizes of CDs bounded and their position in M^4 unaffected, quantum dynamics reduces to a local dynamics inside CDs defining sub-cosmologies. In particular, the age distributions and properties of stars depend only weakly on the value of cosmic time - stars older than the Universe become possible in standard view about time.

4. The flux tube picture about galaxies and larger structures explains the flat velocity spectrum of spiral galaxies if they correspond to tangles of long cosmic strings in which string thickens to flux tube. For elliptical galaxies the cosmic string would be relatively short that the velocity spectrum would not be flat. There would be no dark matter halo except possibly due to the presence of cosmic strings in galactic plane. Some anomalies strongly suggesting the presence of quantum coherence in scales of even billion light years. This could be due to the presence of long quantum coherent structures consisting of flux tube bundles.

5. The presence of dark matter in TGD sense having huge effective Planck constant $h_{eff} = h_{gr} = GMm/\beta_0$ provides a general solution of the well-known angular momentum problem. As the energy associated with thickening cosmic string is transformed to ordinary matter, it must start to rotate around the string to avoid falling back to the cosmic string. This is consistent with angular momentum conservation only if cosmic string generates an opposite angular momentum by developing a helical structure [37] such that dark matter flows along string and rotates at the same time. This solves the well-known angular momentum of the GRT based models.

6. The general model suggests that at least quasars and perhaps all AGNs are actually white-hole like objects as time reversals of blackhole-like objects. The TGD counterpart of BZ model support this view.

3 TGD counterpart of BZ model for central engine

The best manner to explain the TGD variant of BZ model inspired by the Penrose process is to start from the abstract of the original article [10] (`https://cutt.ly/9QtD0kK`).

"It is shown that if the magnetic field and angular momentum of a Kerr blackhole are large enough, the vacuum surrounding the hole is unstable because any stray charged particles will be electrostatically accelerated and will radiate, with the radiation producing electron-positron pairs so freely that the electromagnetic field in the vicinity of the event horizon will become approximately force-free.

Equations governing stationary force-free electromagnetic fields in Kerr spacetime are derived, and it is found that energy and angular momentum can be extracted from a rotating blackhole by a purely electromagnetic mechanism.

The present concepts are applied to a model of an active galactic nucleus containing a massive blackhole surrounded by an accretion disk."

3.1   Beltrami field as the TGD counterpart of force-free field

1. Force free field corresponds in TGD to a generalized Beltrami field conjectured to define very general solutions of field equations [36]. Beltrami fields are analogous to the solutions of Maxwell's equations in the sense that they are dissipation free. This corresponds to the vanishing of the Lorentz force giving $\rho E + j \times B = 0$ and vanishing dissipation giving $j \cdot E = 0$. Therefore the covariant divergence of the energy momentum tensor vanishes and it is plausible that Einstein's equations hold true at the field theory limit.

2. If Beltrami fields correspond to minimal surfaces and at the same time extremals of Kähler action (this requires what I call Hamilton-Jacobi structure as a generalization of ordinary complex structure), the analogs of massless field equations are satisfied as for Maxwell action.

3. The topologies of Beltrami fields are extremely composite and represent knotting and linking of field lines [2, 6, 4, 5]. Helical flux tubes with electric fields parallel to the flux tubes are excellent candidates for Beltrami fields. Configurations of magnetic field, electric field and current such that all three are mutually orthogonal, are suggestive.

   Quantum classical coherence together with the absence of dissipation suggests that supracurrents proportional to j are present.

3.2   Whitehole-like object as the TGD counterpart of Kerr blackhole

Kerr blackhole is used as a model for a rotating blackhole with magnetic field [3, 7] (`https://cutt.ly/sQuKXth`) has besides horizon an outer surface known as ergosphere. At ergosphere light-moving around light-like geodesics is stationary from the point of view of a very distant stationary observer.

In the TGD framework, Kerr blackhole is replaced with a white-hole like object (WH) as a time reversal of a blackhole-like object (BH). This is possible in ZEO (ZEO) [32, 42]. WH is a portion of a tangle of a cosmic string, which has thickened to a flux tube. This reduces string tension as energy per unit length and the system emits energy as ordinary matter to the environment leading to the creation of the galaxy. The process is analogous to the decay of the inflaton field.

3.3   Are secondary objects with quantum coherence scale of order λ_{gr} possible?

All BHs and could be seen as gravitationally quantum coherent objects. For $\beta_0 = 1/2$ quantum gravitational coherence length is at least r_s. If $\beta_0 \leq 1/2$ holds true, r_S is the smallest possible quantum gravitational coherence length.

However, Nottale's hypothesis would assign quantum gravitational coherence length of at least $\lambda_{gr} = GME/v_0$ also to objects which are not BHs. As a matter of fact, TGD leads to a generalization of the notion of BH as a model for the final state of the star [30].

The following intriguing observations inspire the question whether $r_s/2\beta_0$ could define the scale of quantum gravitational coherence also for the secondary objects associated with the astrophysical objects with mass M rather than only WHs and BHs.

1. For Sun λ_{gr} is rather near to the radius of Earth for $\beta_0 \simeq 2^{-11}$ for inner planets. Could this be interpreted in terms of quantum coherent structure formed by parallel flux tubes such that planets correspond to tangles associated with them?

2. The central engines of AGNs have typical size about 2AU, where AU is the distance of Earth from the Sun and size of the inner planetary system [36]. For the blackhole in the center of Milky Way the radius is about .4 AU.

   Is this a mere coincidence or could the important size scales of stellar systems correspond at some level to quantum gravitationally coherent objects with gravitational Compton length determined by AGN? This could be the case if flux tube flux tube bundle from AGN with $h_{eff} = \hbar_{gr}(AGN)$ extends to the galactic plane and forms stellar systems as sub-tangles.

3. One could see the central engine as a provider of metabolic energy to the quantum coherent system formed by the flux tube bundle analogous to a laser beam. The energy feed is necessary to preserve the value of $\hbar_{gr}$ since it tends to be reduced spontaneously. For instance, flux tubes can separate from the bundles as tubes with $h_{eff} = h$. In the terminology of TGD inspired quantum biology, the flux tube bundle would be the magnetic body of the AGN.

3.4 The TGD counterpart of the Penrose process

A popular description of the Penrose process is as follows. Outside the Kerr blackhole a piece of mass split to two pieces. One piece falls to the blackhole and the other piece goes outside (one can imagine that a rocket accelerates outwards and the fuel from a rocket falls into the blackhole). The blackhole piece can be said to have negative energy. By energy conservation this means that one obtains energy from the blackhole. Also the angular momentum of WH is reduced.

One can say that the 3-space around Kerr blackhole is captured into a rotating motion. This applies also to magnetic fields so that the flux lines get twisted and helical patterns are generated. The matter associated with flux lines generates angular momentum.

1. In the TGD framework, the matter from WH forms gravitationally bound states and must rotate in order to avoid falling back to WH. Angular momentum conservation, which is exact in TGD and forces the twisting of the flux tubes so that linear motion forces rotational motion and the system develops classical angular momentum. Also the charged particles flowing along the helical flux tube contribute to the angular momentum. For space-time surfaces, the notion of space-time being captured to rotation is very concretely realized.

2. Also electric field is generated along the helical flux tube to guarantee vanishing of the Lorentz force. In the recent case there is however no accelerated motion. Rather, the flow is dissipation free and gives rise to analog of a supracurrent and/or laser beam. The natural guess for the radius of the helix is as the Schwartschild radius. Quantum gravitational coherence would result from $\hbar_{gr} = GM/\beta_0$ at least in the scale defined by $\lambda_{gr} = r_s/2\beta_0$ equal to Schwartschild radius for $\beta_0 = v_0/c = 1/2$

3.4.1 The magnetic fields associated with the central object and jets

I have considered the identification of BH/WH as a volume filling tangle of a cosmic string thickened to a flux tube and having an ordinary value or relatively small value of h_{eff} much smaller than $\hbar_{gr}$ [30, 31].

The flux tube outside BH/WH could form a flux tube bundle of parallel flux tubes having interpretation as manysheeted space-time with respect to CP_2 or more plausibly $M^2 \times S^2$, serving therefore as the arena of physics instead of M^4. The $\hbar_{gr}/h_0$ would correspond to parallel flux tubes forming a quantum coherent structure.

This helical quantum coherent flux tube structure would be parallel to cosmic string but could have much larger size that the jet length of order 5000 ly. Both would be closed. The tangles of the cosmic string could give give rise to other galaxies, stars, and even planets as suggested [30, 31]. Cyclotron energy in the quantum coherent be however proportional to $\hbar_{gr}$.

In the case of spiral galaxies, this cosmic string would be very long and could connect galaxies: the recent findings suggest that the surrounding structure rotates. This leads to the proposal that the angular momentum of dark matter associated witht this structure compensates for that of ordinary matter [37]: this would solve the well-known angular momentum problem of GRT based cosmology. For elliptical galaxies about which M87 is an example, the velocity spectrum is not flat and the flux tube bundles should close in a relatively short scale.

In the WH/BH region there would be much thinner volume filling flux tube - flux tube spaghetti with relatively small value of h_{eff}. The total flux for a single strand of tangle would be that for the cosmic string itself. There are analogous systems in biology. Proteins consist of parts, which are random coils, helical regions, and planar tangles. These geometries would be induced by the underlying flux tube parallel to the protein. In the recent case, the tangles would be 3-D and have cylindrical geometry in the simplest situation.

Consider now a more detailed model for the magnetic field of M87. The central region is expected to have very strong magnetic field whereas the surrounding region would have much weak magnetic field consisting of parallel flux tubes forming a quantum coherent structure with the length of quantum coherence larger than λ_{gr} identified as parallel space-time sheets but with respect to CP_2 rather than M^4.

Consider first the quantum coherent flux tube structure outside WH surrounding the jet like glove.

1. Suppose that the magnetic field (see the figure of $\mathtt{https://cutt.ly/dQtDK7Q}$) external to the central region corresponds to that for a tangle formed by a cosmic string thickened to a flux tube. $\hbar_{gr}$ would be the number of the parallel flux tubes of the tangle. Many-sheetedness holds true with respect to CP_2 - one can say that CP_2 or $M^2 \times S^2$ serves as the arena of physics instead of M^4.

2. The estimate for the radius of the flux tube obtained by dividing the area πr_s^2 with $\hbar_{gr}/\hbar \simeq 10^{24}$ is of order 2 m so that the helical flux tube structure outside the central object cannot correspond to the observed stripy structure for the magnetic field seen in the polarization patterns. Could the observed stripes correspond to sub-bundles with $\hbar_{eff} \leq \hbar_{gr}$? Or could the quantized monopole flux for the flux tube proportional to the area of the flux tube be large?

3. A reasonable working hypothesis is that the strength of B outside the central region is not more than 100 Gauss in the recent case. Second working hypothesis is that the entire B outside WH is a monopole flux field.

   For $\beta_0 = 1/2$, the gravitational cyclotron energy is $E_{c,gr} \simeq 10^{10}$ GeV: this is by an order of magnitude smaller than the minimal energy scale of ultrahigh energy (UHE) cosmic rays, which is by definition larger than 10^{11} GeV. This supports $\beta_0 = 1/2$ as also the fact that λ_{gr} is equal to r_S, which is a good guess form the minimal value of λ_{gr}

   In the case of M87, quantum coherent emission - whatever it actually means - should take place in a region of size scale about r_s. The length of visible jets is about 5,000 ly Could this scale correspond to the length scale of the helical flux tube tangle in the vertical direction outside the central object?

According to the earlier model [30] the central region corresponds to a volume filling monopole flux tube tangle formed from a cosmic string by thickening and having much narrower flux tubes and by flux conservation with much stronger field strength.

1. The field strength in the central region identified as WH could have considerably higher values. Active galactic nuclei (AGNs) ($\mathtt{https://cutt.ly/DQtFhH3}$), which include quasars, emit jets and the emitted power is typically 10^{38} W to be compared with 3×10^{36} W for M87. The typical radius of the central region is $R \simeq 2$ AU for the active region and about 3 orders of magnitude larger than in the case of M87.

2. The estimate of [9] ($\mathtt{https://cutt.ly/DQtFzD3}$) gives a magnetic field $B \sim 10$ Gauss in the jet region whereas the field strength in the horizon is estimated to be in the range $10^6 - 10^7$ Gauss. Also near-horizon magnetic fields in the range $10^8 - 10^{11}$ Tesla have been proposed for AGNs.

One obtains a rough estimate for the radius r of the flux tube assuming that inside WH the tangle fills the volume.

1. Assume that the particles inside the flux tube are protons with $h_{eff} = h$ and with an average distance equal to the radius of the flux tube. The number N of protons is $N = V/v = (r_s/r)^3$ from which $r = (r_s^3/N)^{1/3}$. One can also express N as M/m_p so that one has $r = ((2G)^3 M^2 m_p)^{1/3} \propto M^{2/3}$ so that r increases with the mass of the central object. For M87 one obtains $r \simeq 10^{-12}$ m, which is of the order electron Compton length. This would suggest that the protons are dark with the value of $h_{eff}/h \simeq m_p/m_e \simeq 2^{11}$. The TGD based model for "cold fusion" leads to the value of h_{eff} [25, 22, 35].

2. By flux conservation, a single unit of flux would correspond to a flux tube with radius equal to Compton length L_c of electron. If L_c equals to the magnetic length $L_B = \sqrt{\hbar/eB}$, one has from $L_B(Tesla) = 26nm$, that the $B \simeq 10^4$ Tesla. If the unit of quantization for magnetic flux is scaled up to $\hbar_{eff}$, the field strength is scaled by $mp/m_e \simeq 2^{11}$ to 2×10^7 Tesla. This option is more natural.

To get some perspective, one can estimate r also for a blackhole with solar mass.

1. From the ratio $M(M87)/M_{Sun} \sim 6.5 \times 10^9$ one obtains that r for a blackhole with solar mass is roughly $.3 \times 10^{-6}$ times smaller than for M87 so that one would have $r \simeq .3 \times 10^{-18}$ m, which is by a factor $.2 \times 10^{-3}$ shorter than proton Compton length 1.3×10^{-15} m.

2. This suggests that the particles are not protons but particles with a higher mass $m = xm_p$. In this case the above estimate is scaled up $r \to x^{1/3}$. p-Adic length scale hypothesis, which is a central element of TGD, indeed predicts scaled up versions of hadron physics assignable to Mersenne primes and their Gaussian counterparts.

3. Ordinary hadron physics would correspond to Mersenne prime $M_{107} = 2^{107} - 1$ M_{89} defines a second candidate for hadron physics with mass scale scaled up by $2^{(107-89)/2} = 2^9$. This would scale up r by a factor $2^{9/3} = 2^3$ to 2.4×10^{-17} m to be compared with the Compton wavelength 2.6×10^{-18} m of M_{89} nucleon so that the interpretation of the flux tubes as having ordinary value of h_{eff} seems to make sense. $\hbar$ would characterize the magnetic body of the BH like object. Note that the model discussed in [30] led to a proposal that the particles at the flux tubes of BH are ordinary nucleons.

4. In this case the magnetic field strength of 2×10^7 Tesla would be scaled by factor 2^{18} to $.5 \times 10^{13}$ Tesla.

3.4.2   The identification of the energy source

Typical radius R of AGN is 2 AU which suggests that the size scale of AGN defines the size of the typical solar system as gravitational Compaton length λ_{gr}. Sagittarius A in the center of Milky way has radius .4 AU, which is the distance of Mercury from Sun. Could solar system be quantum coherent system with respect to Milky Way like Earth with respect to Sun?

One can start from the BZ model [10] of the central object of M87 as a Kerr blackhole. In the TGD framework, blackhole *resp.* whitehole is replaced with a blackhole-like (BH) *resp.* whitehole-like object (WH). WH and BH are time reversals of each other [29, 30, 31].

1. For BZ model, the matter falling into the blackhole would provide the "metabolic" energy feed to the jet and one would have a Penrose process. However, in general relativity the presence of the magnetic field requires Kerr blackhole. However, the arrow of time for Kerr solution is opposite to that of the environment at long distances, which suggests an interpretation as a whitehole having WH rather than BH as its TGD counterpart.

2. WH would have a time direction opposite to that of BH. The energy would come from dark energy and matter of cosmic string (or bundle of cosmic strings) thickening in standard time direction defined by the environment. WH as a quantum coherent object would receive this energy and emit it as jets in opposite directions along the flux tube carrying Beltrami field with electric component. The emission of the energy from thickening cosmic string is analogous to the decay of the inflaton field generating ordinary matter.

   This option looks more plausible and will be considered in the sequel.

In the case of M87 WH, one can estimate the rate of mass loss as the energy radiated as jets. The mass of M87, one would have $M(WH) = 6.5 \times 10^9 M_{Sun}$, $M_{Sun} = 2.2 \times 10^{57}$ J. The mass loss is $dM/dt = 3 \times 10^{36}$ J/s. This gives $dM/dt = -M/\tau$, $\tau = .7 \times 10^{14}$ y. The proportionality $\hbar_{gr} \propto M$ would mean that the rate for the reduction of quantum coherence scale and algebraic complexity defined as $\hbar_{gr}/dt/\hbar_{gr} = 1/\tau$ is slow. This process is analogous to blackhole evaporation as a time reversal for the formation of a blackhole.

One can consider three mechanisms for the energy emission creating a beam along cosmic string possibly thickened to flux tube also outside the WH.

1. The reduction of $\hbar_{gr}$ involves the splitting of the flux tubes from the quantum coherent flux tube bundle as they become ordinary flux tubes so that the number of dark flux tubes decreases. This splitting could give rise to a radiation associated with the jets. This includes synchrotron radiation in $B \sim 100$ Gauss. WH would lose energy as ordinary particles.

2. If the flux tube structure is of size of order of the length of jets about 5000 ly, the emission of closed flux tube bundles by reconnection could be second mechanism. This process would be analogous to the emission of closed flux tubes from the magnetic field of the solar wind. Also now $\hbar_{gr}$ should be reduced.

   The mysterious ultrahigh energy (UHE) cosmic rays could be produced in this process creating the analog of solar wind so that they could be seen as a support for the $\hbar_{gr}$ hypothesis. This process could be regarded as quantum coherent emission with the rate proportional N^2 rather than N, where $N = \hbar_{gr}/h_0$ is the number of flux tubes. The length $L \sim 5000$ ly could be interpreted as the size of the quantum coherent region and would be considerably larger than $\lambda_{gr} \sim 2 \times 10^{-3}$ ly.

3. Dark cyclotron particles and dark cyclotron photons could form an analog of laser beam as a quantum coherent flux tube structure extending to large distances although it forms a closed flux tube. This beam would be quantum analog of solar wind.

   The dark particles could transform in the interactions with ordinary matter to dark particles with a smaller value of $h_{eff}/\hbar_0 = n$ identifiable as the number of flux tubes of the associated sub-bundle. Eventually the beam would decay to ordinary photons with energies which are multiples of E_c. The occurrence of this kind of process in atmosphere could explain the cosmic ray showers due to UHE cosmic rays.

3.5   Nottale's hypothesis and universal cyclotron energies

The non-relativistic approximation for cyclotron energies does not make sense in the recent case so that a relativistically invariant formula $\hbar_{gr} = GME/\beta_0$ [41] is needed.

1. An approximate formula for cyclotron energy in thew relativistic case is obtained from the d'Alembertian equation

$$(E^2 - p_z^2 - m^2 - (p_T - qA)^2)\Psi = \Psi \ . \tag{3.1}$$

The part depending on p_T and vector potential for a constant magnetic field gives the spectrum of non-relativistic harmonic oscillator $n\hbar_{gr}\omega_c$ Hamiltonian multiplied by $2m$. For a given value of n the are $2n+1$ angular momentum eigenstates states with spin in the range $|m| \leq n$. This gives for the energy eigenvalues

$$E_c^2 = m^2 + p_z^2 + n\hbar_{gr}\frac{qB}{m} = m^2 + p_z^2 + n\frac{GM}{\beta_0}qB \ .$$

$$(3.2)$$

p_z satisfies is given by

$$p_z = k\frac{\hbar_{gr}}{L} = \frac{r_s}{L}\frac{m}{2\beta_0} \ .$$

$$(3.3)$$

If the length L of the closed flux tube is of order 5000 ly, $p_z/m = (r_s/L)(1/2\beta_0)$ is very small one neglect p_z^2 and also m^2. This gives

$$E_c = n \times 2\frac{GM}{\beta_0} \ .$$

$$(3.4)$$

The relativistic formula differs from the non-relativistic naive guess only by the factor 2.

2. $\beta_0 = 1/2$ gives $\lambda_{gr} = r_s$ and is therefore favored value in the vicinity of WH/BH. In the case of M87 WH with mass $M = 6.5 \times 19^9 M_{Sun}$ this gives the estimate $E_c = 10^{10}$ GeV, which corresponds to the highest energy for cosmic rays. Could these quanta propagate to Earth and interact with the nuclei of atmosphere to create cosmic ray showers with highest energies about 10^{11} GeV? Note that the proportionality to magnetic field B and $1/v_0$ allows to consider even higher energies.

3.5.1   Two simple models for cyclotron states in generalized Beltrami fields

To get some idea about the solutions of d'Alembertian for a generalized Beltrami field, consider a constant magnetic field at a straight flux tube parallel to the z-axis.

Assume that the constant velocity v corresponds to a rotation around the z-axis. Beltrami condition gives a constant electric field $E = v \times B$ in the radial direction with electric potential $\phi = qvB\rho$. Note that v corresponds to a parameter of Beltrami flow rather than being identified as a single particle operator $v = p_\phi/m = \partial_\phi/\rho m$.

The minimal substitution $p_\mu \to p_\mu - eA_\mu$ in the d'Alembertian gives

$$(E - qBv\rho)^2 - p_z^2 - m^2 - (p_T - qA)^2)\Psi = 0 \ .$$

$$(3.5)$$

One can write the equation in the form analogous to the non-relativistic Schrödinger equation:

$$(H_1 + H_2)\Psi = \frac{1}{2m}(m^2 + p_z^2 - E^2)\Psi \ ,$$

$$H_0 = (p_T - qA)^2)\Psi \ ,$$

$$H_1 = \frac{(qBv)^2}{2m}\rho^2 - \frac{EqBv}{m}\rho \ .$$

$$(3.6)$$

For $v \leq\leq c$, H_1 can be treated as a perturbation. For $H_0 = (p_T - qA)^2$, eigenstates of p_z, L_z and harmonic oscillator Hamiltonian can be solved exactly in the symmetric gauge $A = B \times \rho$, $\rho = (x,y)$.

H_0 reduces to commuting harmonic oscillator Hamiltonians for energy and angular momentum L_z with operators x, ∂_x and y, ∂_y expressed as linear combinations of oscillator operators a^{dagger}, a and $b^\dagger, b$ which commute with each other (https://cutt.ly/iQtFaPR)

ρ^2-term in H_1 can be expressed as bilinear of these operators involving only terms formed from $a^\dagger$ and a *resp.* $b^\dagger$ and b. This term could be included in H_0 and one can hope that Bogoliubov transformation makes it possible to diagonalize the resulting Hamiltonian. ρ is a square root of ρ^2 and this produces problems. Since this term contains also c-number terms from the commutators $[a^{dagger}, a]$ and $[b^{dagger}, b]$, one can expand this term as a power series and treat it perturbatively. One can also use harmonic ordinary oscillator basis for H_0 and treat the situation perturbatively. Note that H_1 commutes with L_z so that one obtains degeneracy of states with respect to L_z also now.

If one had an ordinary gauge invariance, one could consider a gauge in which Kähler gauge potential A is of form $(A_x, A_y) = (0, Bx)$ (https://cutt.ly/iQtFaPR). This situation is of course interesting as such also in the TGD context and could provide a TGD based model for Hall effect.

For $v = (v_x, v_y, v_z)(0, v, 0)$, the electric part of the Kähler gauge potential is $\phi = vBx$. The effective Hamiltonian contains part $H_0 = (p_y - qBx)^2/2m$ and $H_1 = -(E - qvBx)^2/2m$. p_y commutes with both of them so that one can assume eigenstates of p_y. The sum $H_0 + H_1$ reduces to the following form:

$$H_0 + H_1 = \frac{k_1(x - x_0)^2}{2m} - \frac{k_1 x_0^2}{2} \quad ,$$

$$(3.7)$$

where one has

$$k_1 = \hbar^2 \omega_1^2 \;\; , \;\; \omega_1 = \sqrt{1 - v^2}\,\omega_c = \sqrt{1 - v^2}\,\frac{qB}{m} \;\; , \;\; x_0 = \frac{Ev - p_y}{m}\frac{1}{\omega_1} \;\; . \qquad (3.8)$$

The outcome is a harmonic oscillator Hamiltonian for cyclotron energy scale E_c scaled by factor $\sqrt{1 - v^2}$. The origin is shifted from $x = 0$ to $x_0 = (Ev - p_y)qB/m$. Besides this there is a constant term $-\sqrt{1 - v^2}(qB^2 x_0^2/m)/2$.

For $p_y/E = v$ stating that the motion of particles occurs with velocity v, one has $x_0 = 0$ and no shift occurs for energy and the energy spectrum is only scaled by $\sqrt{1 - v^2}$ factor.

3.6 TGD view of jets

There are may questions to be answered.

What is the central engine causing the jets? How does it function? Where does the energy come from? One must also understand the transversal emission: otherwise the jets would be invisible.

BZ proposal is that he jets are naturally in the direction of flux tube along cosmic string of length about 5000 ly. Motion would be along helical orbits and transversal synchrotron radiation would make the jets visible.

There are also questions related to the TGD proposal.

1. Do the synctrotron states as such define the radiation of jet. Both charged dark particles and dark photons indeed have longitudinal momentum along the helix as required by the requirement that they carry angular momentum.

   Are also N-bosons and N-fermions, whose existence is proposed in the TGD based model of dark genetic codes [45, 38] based on dark photons and dark protons, involved.

2. Could also quantum coherent cyclotron transitions give rise to dark photons with $h_{eff} = h_{gr}$ possibly also in directions transversal to the jet? Do all charged particles emit dark cyclotron radiation in synchrony or is this the case only for the charged particles with the same mass m as the fact that

cyclotron frequencies depend on m suggests? Could the huge radiation power be due to quantum coherence: the radiation would be proportional to N^2 instead of N?

3. Could transversal synchrotron radiation with small value of h_{eff} take place for invidual flux tubes or sub-bundles forming a higher level flux tube? The value of $\hbar_{eff}$ would be smaller than $\hbar_{gr}$ for the sub-bundle. Could ordinary synchrotron radiation involve a separation of single flux tube from the condensate and reduction of $\hbar_{gr}$ to $\hbar$?

4. Could the helical flux tube loop also generate dark matter carrying closed flux tubes by reconnection just as occurs in the solar wind of the Sun? These flux tubes could carry cyclotron states and generate synchrotron radiation in transverse directions. This option is not plausible for long cosmic strings.

3.6.1 A model of jets based on Beltrami fields

The helical structure of the magnetic fields and the fact that the generalized Beltrami fields [2, 6, 4, 5] could represent very general preferred extremals in the TGD framework suggest the following picture.

1. The helical structure of the magnetic field is forced by angular momentum conservation in the thickening of a cosmic string to a flux tube liberating energy as matter. Both the classical energy of the helical flux tubes and the motion of particles along the helical flux tube generates angular momentum compensating the angular of matter feeded into environment, which starts to avoid falling into WH.

2. Assume that longitudinal electric field is present inside flux tubes and induced by the rotation of the flux tube and that the classical em field is force-free and therefore a generalized Beltrami field. At single particle level this would mean that Lorenz force vanishes $E = v \times B$.

3. By quantum-classical correspondence, one expects non-dissipative supra current along the helical flux tube. There would be no dissipation by radiation in the ideal situation. One could have analogs of supra currents and laser beams along cosmic string in cosmic scales.

4. The analogs of supracurrents and laser beams would decay to dark particles with a smaller value of $h_{eff}/\hbar_0$ identifiable as the number of flux tubes of the associated sub-bundles. Eventually the beam would decay to ordinary photons. This kind of process occurring in atmosphere could explain ultra-high energy cosmic rays. This process could involve reconnection.

   The transversal synchrotron radiation could be created as $h_{eff} = h_{gr}$ for a particle decreases and photon with much smaller value of h_{eff} leaks out from the dark flux tube?

5. The quantum coherence along jet decreases gradually if its h_{eff} decreases. The WH associated defining a tangle of the cosmic string would however serve as a source of metabolic energy so that the value of $\hbar_{gr}$ would not be reduced. The situation would be very much like in TGD inspired biology except that the life in question would be in cosmological scales. The value of $\hbar_{gr}$ for AGNs would be huge as compared for that for Earth based life.

There is clearly an analog with the evaporation of blackholes and one can ask whether blackhole evaporation is induced by the change of the arrow of time for BH transforming it to WH.

3.6.2 How do the jets shine?

Jets shine, which means that photons and other particles are emitted in directions nearly parallel to the particle motion. One expects synchrotron radiation - possibly dark but with relatively small h_{eff}- in the plane orthogonal to the flux tube. According to the BZ proposal, particles rotating along the helical flux tubes in the direction of jet axis could emit the radiation as synchrotron radiation in directions roughly tangential to the helical orbit. How could this mechanism be realized in the TGD framework?

The synchrotron radiation could be seen as a leakage of particles from the helical flux tubes occurring directions near the tangential direction. This process could relate closely to the reduction of quantum coherence scale as flux tubes are separated in the quantum coherent flux tube bundle to a flux tube with ordinary value of Planck constant. For $\hbar_{gr}/\hbar \sim 8.1 \times 10^{24}$ (electron) and area πr_s^2 of the cylinder determined by $r_s \sim 19.5 \times 10^9$ km, the transversal area of single flux tube would be about 2.4π m^2. Cyclotron energy scale 5×10^{-7} eV corresponds to 2 meter wavelength for a radiowave photon.

One could also consider a splitting to quantum coherent sub-bundles of flux tubes with h_{eff} proportional to the number n of split flux tubes and cyclotron frequency scaled up accordingly.

3.6.3 UHE cosmic rays as a quantum gravitational effect?

The origin of ultra high energy (UHE) cosmic rays is poorly understood in the standard physics framework. The reader can consult a Wikipedia article about UHE cosmic rays (`https://cutt.ly/7QtFc7h`) and there is also Quanta Magazine article about the topic (`https://cutt.ly/ZQtFnff`).

Could the huge dark cyclotron energies for quantum gravitational cyclotron states make it possible to understand the origin of UHE cosmic rays? The following proposal is perhaps the simplest mechanism that one can imagine in TGD framework.

Assume that the monopole part $B_{end}(M87)$ of the magnetic field of M87 WH is equal to the magnetic field $B(M87) \sim 100$ Gauss: $B_{end}(M87) = B(M87)$.

For electron with $h_{eff} = h$ in $B_{end} = .2$ Gauss associated to Earth by the model for the findings of Blackman and others [12] gives $E_c(e) = 4.96 \times 10^{-7}$ eV. For $B_{end}(M87) = B(M87) \simeq 100$ Gauss $B_{end} = .2$ Gauss is scaled up by factor 500. $\hbar$ is scaled up by $\hbar_{gr}/\hbar \simeq 8.1 \times 10^{24}$. The relativistic formula $E_c(M87) = 2GMQB/\beta_0$ for the cyclotron energy scales gives for $\beta_0 = 1/2$ $E_c(M87) = 2(\hbar_{gr}/\hbar)(B_{end}) \simeq 10^{10}$ GeV . This is extremely high energy: note that 10^{10} GeV corresponds to the lower bound for UHE cosmic rays with energies of order 10^{11} GeV.

All charged particles have this energy scale independently of their mass. What could happen to dark photons emitted in cyclotron transitions and dark charged particles with energy about 10^{10} GeV if they are emitted from WH?

Is there any separate emission process or do the dark particles in cyclotron states travel along the direction of jet as an analog of laser beam?

1. If the laser beam option is realized, dark cyclotron photons and particles (this includes charged particles with large range of masses) could travel to Earth as such. The interaction with the atomosphere would make this dark cosmic ray observable. This process would involve reduction of $\hbar_{gr}$ to $\hbar$.

2. Is the direct transformation to single photon possible as in the case of the decay of dark photons to biophotons? This would conform with the interpretation as ordinary UHE cosmic ray - say proton.

3. TGD suggest also the decay to large number of photons with smaller values of h_{gr} and the flux tube bundle picture suggests that the sum of the integers n_i characterizing $h_{eff,i}$ and the number of flux tubes in the sub-bundle associated satisfies $\sum n_i \simeq \hbar_{gr}/\hbar_0$. For $n_i = 1$ the outcome would be a bunch of ordinary radiowave photons with $E_c \simeq 5 \times 10^{-7}$ eV. A natural expectation is that the decay process occurs as a cascade in which sub-bundles decay to smaller sub-bundles or transform to single ordinary photon with energy $n_i E_c$. Irrespective of the details of the decay process, the total energy of the cosmic ray show would correspond to $8.1 \times 10^{24} E_c$.

4. If the formula from the d'Alembertian is a good approximation, most of the energy of dark cyclotron particle parallel to the flux tube bundle, in particular the energy of photon, would be in transversal degrees of freedom in cyclotron motion at flux tube although the particle is by its huge energy massless in an excellent approximation.

ISSN: 2153-8301   Prespacetime Journal   www.prespacetime.com
Published by QuantumDream, Inc.

(Published in Prespacetime Journal | December 2021 | Volume 12 | Issue 4 | pp. 429-448)　**120**
Pitkänen, M., *TGD View of the Engine Powering Jets from Active Galactic Nuclei*

The synchrotron particles would propagate very slowly in the direction of the jet although the are massless: $p_z/E \le\le 1$. In fact, TGD based view about particle massivation relies on zitterbewung so that all particles are predicted to massless in short enough scales. This does not conform with the properties of MEs for which the longitudinal momentum is light-like although transversal degrees of freedom are present.

Received August 8, 2021; Accepted December 4, 2021

References

[1] Zeeman EC. *Catastrophe Theory*. Addison-Wessley Publishing Company, 1977.

[2] Lakthakia A. *Beltrami Fields in Chiral Media*, volume 2. World Scientific, Singapore, 1994.

[3] Heinicke C and Hehl FW. Schwarzschild and Kerr solutions of Einstein's field equation: An Introduction. *Int J. Mod Phys. D*.

[4] Ghrist R Etnyre J. An index for closed orbits in Beltrami field, 2001. Available at: `http://arxiv.org/abs/math/010109`.

[5] Marsh GE. *Helicity and Electromagnetic Field Topology*. World Scientific, 1995.

[6] Bogoyavlenskij OI. Exact unsteady solutions to the Navier-Stokes equations and viscous MHD equations. *Phys Lett A*, pages 281–286, 2003.

[7] Adamo T and Newman ET. The Kerr-Newman metric: A Review. *Scholarpedia*, 9(10), 2014. Available at:`http://arxiv.org/abs/1410.6626`.

[8] Nottale L Da Rocha D. Gravitational Structure Formation in Scale Relativity, 2003. Available at: `http://arxiv.org/abs/astro-ph/0310036`.

[9] Silant'ev NA et al. Magnetic fields of active galactic nuclei and quasars with polarized broad Hα lines, 2012. Available at: `https://arxiv.org/pdf/1203.2763.pdf`.

[10] Blandford RD and Znajek RL. Electromagnetic extraction of energy from Kerr black holes. *Monthly Notices of the Royal Astronomical Society*, 179:433–456, 1977. Available at: `https://doi.org/10.1093/mnras/179.3.433`.

[11] Mead LR Ringermacher HI. Observation of Discrete Oscillations in a Model-independent Plot of Cosmological Scale Factor vs. Lookback Time and a Scalar Field Model, 2015. Available at: `http://arxiv.org/abs/1502.06140`.

[12] Blackman CF. *Effect of Electrical and Magnetic Fields on the Nervous System*, pages 331–355. Plenum, New York, 1994.

[13] Pitkänen M. About Strange Effects Related to Rotating Magnetic Systems . In *TGD and Fringe Physics*. Available at: `http://tgdtheory.fi/pdfpool/Faraday.pdf`, 2006.

[14] Pitkänen M. Quantum Model for Bio-Superconductivity: I. In *TGD and EEG*. Available at: `http://tgdtheory.fi/pdfpool/biosupercondI.pdf`, 2006.

[15] Pitkänen M. Quantum Model for Bio-Superconductivity: II. In *TGD and EEG*. Available at: `http://tgdtheory.fi/pdfpool/biosupercondII.pdf`, 2006.

[16] Pitkänen M. About Preferred Extremals of Kähler Action. In *Physics in Many-Sheeted Space-Time: Part I*. Available at: `http://tgdtheory.fi/pdfpool/prext.pdf`, 2019.

[17] Pitkänen M. About twistor lift of TGD? In *Towards M-Matrix: Part II*. Available at: `http://tgdtheory.fi/pdfpool/hgrtwistor.pdf`, 2019.

[18] Pitkänen M. More about TGD Inspired Cosmology. In *Physics in Many-Sheeted Space-Time: Part II*. Available at: `http://tgdtheory.fi/pdfpool/cosmomore.pdf`, 2019.

[19] Pitkänen M. Some questions related to the twistor lift of TGD. In *Towards M-Matrix: Part II*. Available at: `http://tgdtheory.fi/pdfpool/twistquestions.pdf`, 2019.

[20] Pitkänen M. TGD View about Coupling Constant Evolution. In *Towards M-Matrix: Part I*. Available at: `http://tgdtheory.fi/pdfpool/ccevolution.pdf`, 2019.

[21] Pitkänen M. The Recent View about Twistorialization in TGD Framework. In *Towards M-Matrix: Part II*. Available at: `http://tgdtheory.fi/pdfpool/smatrix.pdf`, 2019.

[22] Pitkänen M. Cold Fusion Again . Available at: `http://tgdtheory.fi/public_html/articles/cfagain.pdf.`, 2015.

[23] Pitkänen M. Hydrinos again. Available at: `http://tgdtheory.fi/public_html/articles/Millsagain.pdf.`, 2016.

[24] Pitkänen M. Strong support for TGD based model of cold fusion from the recent article of Holmlid and Kotzias. Available at: `http://tgdtheory.fi/public_html/articles/holmilidnew.pdf.`, 2016.

[25] Pitkänen M. Cold fusion, low energy nuclear reactions, or dark nuclear synthesis? Available at: `http://tgdtheory.fi/public_html/articles/krivit.pdf.`, 2017.

[26] Pitkänen M. Philosophy of Adelic Physics. In *Trends and Mathematical Methods in Interdisciplinary Mathematical Sciences*, pages 241–319. Springer.Available at: `https://link.springer.com/chapter/10.1007/978-3-319-55612-3_11`, 2017.

[27] Pitkänen M. Philosophy of Adelic Physics. Available at: `http://tgdtheory.fi/public_html/articles/adelephysics.pdf.`, 2017.

[28] Pitkänen M. Expanding Earth Model and Pre-Cambrian Evolution of Continents, Climate, and Life. Available at: `http://tgdtheory.fi/public_html/articles/expearth.pdf.`, 2018.

[29] Pitkänen M. TGD view about quasars. Available at: `http://tgdtheory.fi/public_html/articles/meco.pdf.`, 2018.

[30] Pitkänen M. Cosmic string model for the formation of galaxies and stars. Available at: `http://tgdtheory.fi/public_html/articles/galaxystars.pdf.`, 2019.

[31] Pitkänen M. Solar Metallicity Problem from TGD Perspective. Available at: `http://tgdtheory.fi/public_html/articles/darkcore.pdf.`, 2019.

[32] Pitkänen M. Some comments related to Zero Energy Ontology (ZEO). Available at: `http://tgdtheory.fi/public_html/articles/zeoquestions.pdf.`, 2019.

[33] Pitkänen M. A critical re-examination of $M^8 - H$ duality hypothesis: part I. Available at: `http://tgdtheory.fi/public_html/articles/M8H1.pdf.`, 2020.

[34] Pitkänen M. A critical re-examination of $M^8 - H$ duality hypothesis: part II. Available at: `http://tgdtheory.fi/public_html/articles/M8H2.pdf.`, 2020.

[35] Pitkänen M. Could TGD provide new solutions to the energy problem? Available at: `http://tgdtheory.fi/public_html/articles/proposal.pdf.`, 2020.

[36] Pitkänen M. Comparing the Berry phase model of super-conductivity with the TGD based model. `https://tgdtheory.fi/public_html/articles/SCBerryTGD.pdf.`, 2021.

[37] Pitkänen M. Cosmic spinning filaments that are too long. `https://tgdtheory.fi/public_html/articles/spincstring.pdf.`, 2021.

[38] Pitkänen M. Is genetic code part of fundamental physics in TGD framework? Available at: `https://tgdtheory.fi/public_html/articles/TIH.pdf.`, 2021.

[39] Pitkänen M. Is $M^8 - H$ duality consistent with Fourier analysis at the level of $M^4 \times CP_2$? `https://tgdtheory.fi/public_html/articles/M8Hperiodic.pdf.`, 2021.

[40] Pitkänen M. Negentropy Maximization Principle and Second Law. Available at: `https://tgdtheory.fi/public_html/articles/nmpsecondlaw.pdf.`, 2021.

[41] Pitkänen M. Questions about coupling constant evolution. `https://tgdtheory.fi/public_html/articles/ccheff.pdf.`, 2021.

[42] Pitkänen M. Some questions concerning zero energy ontology. `https://tgdtheory.fi/public_html/articles/zeonew.pdf.`, 2021.

[43] Pitkänen M. Updated version of Expanding Earth model. `https://tgdtheory.fi/public_html/articles/expearth2021.pdf.`, 2021.

[44] Pitkänen M. What could 2-D minimal surfaces teach about TGD? `https://tgdtheory.fi/public_html/articles/minimal.pdf.`, 2021.

[45] Pitkänen M and Rastmanesh R. The based view about dark matter at the level of molecular biology. Available at: `http://tgdtheory.fi/public_html/articles/darkchemi.pdf.`, 2020.

ISSN: 2153-8301   Prespacetime Journal   www.prespacetime.com
Published by QuantumDream, Inc.

Article

Empirical Support for the Expanding Earth Model & TGD View about Classical Gauge Fields

Matti Pitkänen [1]

Abstract

In this article I will discuss some empirical facts providing further support for the Expanding Earth Model (EEM) and TGD view about classical fields. The first strange finding is the large fluctuations of oxygen levels during the Cambrian Explosion. The general form of EEM applies to all astrophysical objects and could explain the strange lack of craters and volcanic activity in Venus suggesting a global resurfacing for 750 million years ago. Contrary to expectations, the magnetic field of Venus vanishes. The TGD based view about gauge fields differs from the standard view in that it allows the notion of monopole flux. The monopole part field would be analogous to the external magnetic field H inducing magnetization M as the non-monopole part of B. Venus would be a perfect diamagnet and even a superconductor whereas Earth would be a paramagnet. In the TGD framework, superconductivity driven by the thermal energy feed from the interior of Venus would be possible. The interior of Venus could be a living system but in a very different sense than Earth.

1   Introduction

In this article I will discuss some empirical facts providing further support for the Expanding Earth Model (EEM) and TGD view about classical fields.

1.1   Findings supporting the Expanding Earth Model

There are two findings that provide further support for EEM and its generalization.

1. The Great Oxidation event that culminated during the Cambrian Explosion involved strong fluctuations of the oxidation level manifesting itself as extinctions and emergence of new species [3]. The expansion of Earth would have involved breakages of the crust bringing oxygen rich water from the Earth interior containing multicellulars to the surface and the dilution of this oxygen rich water would have led to the extinction.

2. The fact that the surface of Venus involves very few craters and the volcanic activity is absent suggests that Venus has effectively turned itself inside out about 750 million years ago [1, 2]. The general form of the EEM applies to all astrophysical objects: could Expanding Venus Hypothesis explain the strange absence of craters and volcanic activity in Venus?

1.2   Why does the magnetic field of Venus vanish?

The TGD based view about gauge fields differs from the standard view in that it allows the notion of monopole fluxes requiring non-trivial space-time topology. The monopole part field would be analogous to the external magnetic field H inducing magnetization M as the non-monopole part of B. Venus would be a perfect diamagnet and even a superconductor whereas Earth would be a paramagnet. In the TGD framework, superconductivity driven by the thermal energy feed from the interior of Venus would be possible.

---

[1]Correspondence: Matti Pitkänen http://tgdtheory.fi/. Address: Rinnekatu 2-4 8A, 03620, Karkkila, Finland. Email: matpitka6@gmail.com.

It has been known for a long time that Venus has a very weak magnetic field although the dynamo model would suggest its presence. TGD based model for the maintenance of magnetic field assumes that the monopole part of field, which requires no currents to maintain it, is analogous to the external magnetic field H, which induces magnetization M ($B = H + M$) suggests that Venus is diamagnetic ($B = 0$) Earth would be paramagnetic.

Superconductors are ideal diamagnets: could Venus be a superconductor? This is possible in the TGD framework. The heat flow from the core of Venus would provide the "metabolic energy feed" making possible the large value of h_{eff} required by high Tc superconductivity above critical temperature [7]. The magnetization created by supercurrents would cancel the external magnetic field: this would provide a TGD based view about the Meissner effect. The interior of Venus could be a living system but in a very different sense than Earth.

2 Cambrian explosion, the Great Oxidation Event, and EEM

I encountered two interesting articles related to the Great Oxidation Event that started long before the Cambrian Explosion (CE) and reached its climax during CE (about 541 million years ago) leading to the oxygen based multicellular life in a very rapid time scale.

The standard view is that oceans before CE had very low oxygen content. The emergence of photosynthesizing cyanobacteria producing oxygen as a side product led to the oxygenation of the atmosphere and to mysteriously rapid evolution of life. How this is possible at all is not understood.

The first popular article (`https://cutt.ly/UQWZA31`) discusses the proposal [3] that the slowing down of the spinning of Earth was somehow related to this. The idea is that the lengthening of the day made photosynthesis by cyanobacteria more effective since their reaction to the dawn of the day was slow. The second article in Quanta Magazine (`https://cutt.ly/PQWZDzD`) tells about the finding [I1] that during the Cambrian Explosion (`https://cutt.ly/1QWZF4E`) the oxygen content of the studied shallow ocean show fluctuations with with about 4-5 peaks. The reduction/increase of the oxygen content was even 40 per cent, which is a huge number. The reduction of oxygen content caused extinctions and its increase was accompanied by the emergence of new species. The mystery is how this could happen so fast and which caused the fluctuations.

2.1 Expanding Earth model very briefly

EEM is not originally TGD based but TGD provides its realization. The proposal is that the Cambrian Explosion was caused by a rapid increase of the radius of Earth by factor 2 [6, 5, 9].

This hypothesis also solves one of the basic mysteries of cosmology. Astrophysical objects participate in cosmological expansion by comoving with it but do not expand themselves. Why? The prediction that the expansion of the astrophysical objects did not occur smoothly but as rapid phase transitions and the expansion was very slow in the intermediate states. Cambrian Explosion would correspond to one particular jerk of this kind in which the radius of Earth grew by a factor 2 (p-adic length scale hypothesis). The length of the day increased by factor 4 from conservation of angular momentum. This might relate to the conjecture of the first article.

The rapid expansion led to the breakage of the Earth crust and to the birth of plate tectonics. It also led to the burst of underground oceans to the surface of the Earth. The photosynthesizing multicellular life had developed in these oceans and emerged almost instantaneously and led to a rapid oxygenation of the atmosphere. One can say that life evolved in the womb of Mother Gaia shielded from meteorites and cosmic rays. No superfast evolution was needed. Already Charles Darwin realized that the sudden appearance of trilobites was a heavy objection against the theory of natural selection.

Possible scenarios for the phase transition are discussed in [9]. The thickening of magnetic flux tubes for water blobs at the surface of Earth led to the increase of the volume of water blobs and induced the increase of h_{eff} a factor 2 for valence electrons but not for the inner electrons. Since valence electrons are

ISSN: 2153-8301

responsible for chemistry, atoms became effectively dark and the water blobs could leak to the interior of Earth. By their darkness they could have much lower temperature and pressure than the matter around them and life could evolve.

2.2   How was photosynthesis possible underground?

What made photosynthesis possible in the underground oceans? One possible explanation is that the photons from the Sun propagated along flux tubes of the "endogenous" part of the Earth's magnetic field as dark photons with $h_{eff} = nh_0 > h$. Endogenous part would be the part of Earth's magnetic field with a strength about 2/5 of the Earth's magnetic field for which flux tubes carry monopole flux: this is possible in TGD but not in Maxwell's theory.

Since these photons behave like dark matter with respect to the ordinary matter, they were not absorbed considerably and reached the water blobs (or actually their magnetic bodies consisting of flux tubes) in underground oceans having a portion with the same value of $h_{eff} \geq h$. Of course, several values of h_{eff} were possible since this is the case in quantum critical system (large values of h_{eff} characterize the quantum scales of long range fluctuations). One can also consider other variants of the model. The ordinary matter in Earth's crust had $h_{eff} = h/2$ and photons with $h_{eff} = h$ propagated to the interior and reached the water blobs with $h_{eff} = h$.

2.3   The sudden emergence of multicellulars and oxygen fluctuations

Before the expansion period was much like the surface of Mars now and contained no oceans, perhaps some ponds allowing primitive monocellular lifeforms. As the ground of Earth broke here and there during the rapid expansion period, lakes and oceans were formed at the surface of Earth. The multicellulars bursted to these oceans and oxygenation of the atmosphere started locally.

Since the oxygen rich water was mixed with the water in the shallow oceans, the local oxygen content of the burst water was reduced and this led to an eventual extinction of many multicellulars in the burst. Burgess Shale fauna contained entire classes, which suffered extinction. In the average sense the oxygen concentration increased and led to the apparent very rapid evolution of multicellulars, which had actually already occurred underground. Of course, also evolution at the surface of Earth took place.

3   Has venus turned itself inside-out and why its magnetic field vanishes?

News about unexpected findings relating to the physics of astrophysical objects emerge on an almost daily basis. The most recent news (`https://cutt.ly/YQSZgpv`) told about the lack of craters and volcanic activity in Venus (`https://cutt.ly/wQSZzaS`). The findings are actually not new. The resurfacing history of Venus was summarized 1979 by Schaber et al [1] . Turcotte and Rome have proposed cyclic global catastrophic events as an analog of the plate tectonics allowing a heat transfer from the interior of Venus and effectively turning Venus inside out [2].

The Venus does not have appreciable magnetic field although dynamo mechanism suggests magnetic field as in the case of Earth, has been also known.

3.1   Has Venus turned itself inside-out?

The surface of Venus was expected to have craters, just like the surface of Earth, Moon, and Mars but the number of craters is very small. The surface of Venus also has weird features and many volcanoes. Also trace signs of erosion and tectonic shifts were found. The impression is that the surface of Venus had been turned inside out in a catastrophic event that occurred about 750 million years ago.

Since Venus is our sister planet with almost the same mass and radius, it is interesting to notice that the biology of Earth experienced the Cambrian explosion 541 million years ago.

1. The TGD explanation for Cambrian Explosion relies on EEM [6, 5, 9]. The model assumes that there was a relatively fast increase of the Earth's radius by factor, which led to the burst of underground oceans to the surface of the Earth and led to the formation of oceans. Standard cosmology predicts a continuous smooth expansion of astrophysical objects. Contrary to this prediction, astrophysical objects do not seem to expand smoothly. In the TGD Universe, the smooth expansion is replaced by rapid jerks and the Cambrian Explosion would be associated with this kind of phase transitions.

2. In this expansion the multicellular photosynthesizing life burst to the surface. This explains the sudden emergence of highly evolved life forms during the Cambrian Explosion that Darwin realized to be a heavy objection against his theory.

3. There are many objections to be circumvented. For instance, how photosynthesis could evolve in the underground ocean. Here TGD views dark matter as $h_{eff} = nh_0$ phases of ordinary matter, which are relatively dark with respect to each other, come in rescue. Dark water blobs could leak into the interior of Earth and the solar light possessing a dark portion could do the same so that photosynthesis became possible [9].

4. Did Venus experience a similar rapid expansion 200 million years earlier, about 750 million years ago (or maybe roughly at the same time). Venus does not have water at its surface. This can be understood in terms of heat from solar radiation forcing the evaporation of water and subsequent loss. This also prevented the leakage of the water to the interior of Venus. If there were no water reservoirs inside Venus, no oceans were formed. The cracks of the crust created expanding areas of magma, which were like the bottoms of the oceans at Earth. Also at Earth a fraction about 2/3 of the Earth's surface is sea bottom.

3.2   Why does Venus not possess a magnetic field?

Venus also offers a second puzzle. Venus does not have an appreciable magnetic field although it has been speculated that it has had it (`https://cutt.ly/VQSZt9m`). The solar dynamo mechanism would suggest its presence.

1. TGD predicts that there are two kinds of flux tubes carrying Earth's magnetic field B_E with a nominal value of .5 Gauss. This applies quite generally. The flux tubes have a closed cross section - this is possible only in TGD Universe, where the space-time is 4-surface in $M^4 \times CP_2$. The flux tubes can have a vanishing Kähler magnetic flux or non-vanishing quantized monopole flux: this has no counterpart in Maxwellian electrodynamics.

   For Earth, the monopole part would correspond to about .2 Gauss - 2/5 of the full strength of B_E.

2. Monopole part needs no currents to maintain it and this makes it possible to understand how the Earth's magnetic field has not disappeared a long time ago. This also explains the existence of magnetic fields in cosmological scales.

   The orientation of the Earth's magnetic field is varying. In the TGD based model the monopole part plays the role of master. When the non-monopole part becomes too weak, the magnetic body defined by the monopole part changes its orientation. This induced currents refresh the non-monopole part [4]. The standard dynamo model is part of this model.

3. There is an interesting (perhaps more than) analogy with the standard phenomenological description of magnetism in condensed matter. One has $B = H + M$. H field is analogous to the monopole part and the non-monopole part is analogous to the magnetization M induced by H. $B = H + M$

would represent the total field. If this description corresponds to the presence of two kinds of flux tubes, the TGD view about magnetic fields would have been part of electromagnetism from the beginning!

Flux tubes can also carry electric fields and also for them this kind of decomposition makes sense. Could also the fields D, P, and E have a similar interpretation?

In the linear model of magnetism, one has $M = \chi H$ and $B = \mu H = (1 + \chi)H$. For diamagnets one has $\chi \leq 0$ and for paramagnets $\chi \geq 0$. Earth would be paramagnetic with $\chi \simeq 3/2$ if the linear model works. χ is a tensor in the general case so that B and H can have different directions.

4. All stars and planets, also Venus, correspond to flux tube tangles formed from monopole flux tubes. This leaves only one possibility. Venus behaves like a super-conductor and is an ideal diamagnet with $\chi = -1$ so that B vanishes. The monopole part would be present however.

   This could provide a totally new insight to the Meissner effect and loss of superconductivity. In TGD the based model [7], monopole flux tubes carry supracurrent. The BCS model however requires the absence of a magnetic field. Could the induced non-monopole field cancelling the monopole part. Venus would indeed be a superconductor!

5. The TGD based model of superconductivity [7] also predicts superconductivity driven by an external energy feed would be also above critical temperature. The energy feed would increase the value of h_{eff} and below the critical temperature it would be provided by the energy liberated in the formation of Cooper pairs, which need not actually be the current carriers since dark electrons can carry the current without dissipation. In TGD inspired biology and quite universally, the basic role of metabolic energy feed is to prevent the reductions of the values of h_{eff}.

   Superconductivity means in the TGD framework large h_{eff} and therefore complexity, intelligence, and long quantum coherence length [10]. Could Venus be alive but in a very different sense than Earth?

6. Could the superconductivity be forced by the thermal energy feed from the interior of Venus? The tilt of the rotation axis relative to the plane of rotation around the Sun is very small for Venus, about 3 degrees and much smaller than for the Earth. This implies that the surface temperature of Venus is roughly constant. At Earth plate tectonics makes possible the heat transfer from the interior to the surface and its leakage to outer space. For Venus this is not possible. Could the energy flow from the interior of Venus force the superconductivity by increasing the values of h_{eff}. This would in turn force the vanishing of the magnetic field of Venus.

7. Sun has an enormous feed of metabolic energy from the core: could it be alive? Also in the case of Earth, the energy feed from the interior could have been crucial for the development of life in the interior of Earth and made possible even the development of photosynthesis.

   The possibility that life actually appears in cosmic scales and is associated with quantum coherent flux tube networks associated with the active galactic nuclei usually identified as supermassive blackholes containing stellar and planetary systems as tangles is suggested by the TGD based model of galactic jets [8] explaining also ultrahigh energy cosmic rays. The model inspires the proposal that active galactic nuclei having typically sizes 1-2 AU (!) involve gravitationally quantum coherent regions of radius at most of the Schwartschild radius defining a minimal gravitational Compton length [8].

8. Also Mars lacks the global magnetic field although it has auroras assigned with local fields. Could also Mars be alive in the same same sense as Venus? Note that the recent radius of Mars is about $1/2$ of Earth's radius. If Venus expanded by factor 2, all these 3 planets would have had roughly the same radius for about 750 million years ago. Mars would be waiting for the moment of expansion.

Received August 14, 2021; Accepted December 4, 2021

References

[1] Schaber R et al. The Resurfacing History of Venus, 1997. Available at: `https://www.researchgate.net/publication/234304451_The_Resurfacing_History_of_Venus`.

[2] Romeo I and Turcotte DL. Resurfacing on Venus . *Planetary and Space Science*, 58(10):1374–1380, 2010. Available at: `https://doi.org/10.1016/j.pss.2010.05.022`.

[I1] He T et al. Possible links between extreme oxygen perturbations and the Cambrian radiation of animals. *Nature Geoscience*, 12:468–474, 2019. Available at: `https://www.nature.com/articles/s41561-019-0357-z`.

[3] Klatt JM et al. Possible link between Earth's rotation rate and oxygenation. *Nature Geoscience*, 14:564–570, 2021. Available at: `https://www.nature.com/articles/s41561-021-00784-3`.

[4] Pitkänen M. Maintenance problem for Earth's magnetic field. Available at: `http://tgdtheory.fi/public_html/articles/Bmaintenance.pdf.`, 2015.

[5] Pitkänen M. Expanding Earth hypothesis, Platonic solids, and plate tectonics as a symplectic flow. Available at: `http://tgdtheory.fi/public_html/articles/platoplate.pdf.`, 2018.

[6] Pitkänen M. Expanding Earth Model and Pre-Cambrian Evolution of Continents, Climate, and Life. Available at: `http://tgdtheory.fi/public_html/articles/expearth.pdf.`, 2018.

[7] Pitkänen M. Comparing the Berry phase model of super-conductivity with the TGD based model. `https://tgdtheory.fi/public_html/articles/SCBerryTGD.pdf.`, 2021.

[8] Pitkänen M. TGD view of the engine powering jets from active galactic nuclei. `https://tgdtheory.fi/public_html/articles/galjets.pdf.`, 2021.

[9] Pitkänen M. Updated version of Expanding Earth model. `https://tgdtheory.fi/public_html/articles/expearth2021.pdf.`, 2021.

[10] Pitkänen M and Rastmanesh R. The based view about dark matter at the level of molecular biology. Available at: `http://tgdtheory.fi/public_html/articles/darkchemi.pdf.`, 2020.